CLARENDON LECTURES IN GEOGRAPHY AND ENVIRONMENTAL STUDIES

General Editor

Gordon Clark

Geography and Economy
Three Lectures

———

ALLEN J. SCOTT

CLARENDON PRESS · OXFORD

OXFORD

UNIVERSITY PRESS

Great Clarendon Street, Oxford ox2 6DP

Oxford University Press is a department of the University of Oxford.
It furthers the University's objective of excellent in research, scholarship,
and education by publishing worldwide in

Oxford New York

Auckland Cape Town Dar es Salaam Hong Kong Karachi
Kuala Lumpur Madrid Melbourne Mexico City Nairobi
New Delhi Shanghai Taipei Toronto

With offices in

Argentina Austria Brazil Chile Czech Republic France Greece
Guatemala Hungary Italy Japan Poland Portugal Singapore
South Korea Switzerland Thailand Turkey Ukraine Vietnam

Oxford is a registered trade mark of Oxford University Press
in the UK and in certain other countries

Published in the United States
by Oxford University Press Inc., New York

British Library Cataloguing in Publication Data

Data available

Library of Congress Cataloging-in-Publication Data
Scott, Allen John.
Geography and economy: three lectures/Allen J. Scott. p.cm
ISBN 0-19-928430-X (alk. paper)
1. Economic geography. I. Title.
HF 1025.S362 2006
330.9—dc22
2005025981

Typeset by SPI Publisher Services, Pondicherry, India
Printed in Great Britain
on acid-free paper by
Biddles Ltd., King's Lynn

ISBN 0-19-928430-X 978-0-19-928430-6

1 3 5 7 9 10 8 6 4 2

PROEM

The three chapters that follow form the basis of the Clarendon Lectures in Geography and Environmental Studies that I was privileged to present at the School of Geography and the Environment, Oxford University, over the period 4–6 May, 2005.

Since I have never before been involved in a public lecture series quite as ambitious as this one, I was at the outset puzzled about what the subject of these addresses should be, and in what style I should present them. Should they focus on cutting-edge research results? Should they be pedagogic, like glorified undergraduate lectures? Should they somehow represent my own attempt to come to terms with the debates that continually rage throughout the field of economic geography? In the end, I have settled for a compromise solution that attempts to combine something of all three approaches while at the same time allowing myself the luxury of trying to re-express and update a number of theoretical ideas that have preoccupied me over the last couple of decades.

Geography and Economy is not a book that attempts to explore the entire terrain opened up by the ambitious programmatic promise that its title may seem to signify. I have sought, rather, to focus the argument on what I take to be some burning theoretical and practical questions at the present time, and to explore these at the juncture where geography and economy meet. The concomitant points of emphasis, each of which occupies a chapter in what follows, revolve around

(a) the division of labour and the ways in which it intertwines with locational outcomes at every scale of analysis;
(b) the creative field, which is to be understood as a grid of spatial relationships that functions as a powerful stimulus of entrepreneurship and innovation; and
(c) the regional bases of economic take-off and development.

These three sets of issues, all of which are intimately interrelated, strike me as providing the strong points from which so many of the other big issues of modern economic geography can be fruitfully attacked. They are certainly fundamental to any effort to decipher the structure and dynamics of the economic landscape, and how the landscape, in its turn, moulds the temper and performance of capitalism in different places.

To begin with, the division of labour is one of the primary factors underlying the spatial differentiation of modern society. An older school of thought in geography was much concerned with the physical and environmental conditions giving rise to variations in modes of life from one place to another. In contemporary society, these conditions still play an important role. Increasingly, however, they are being overridden by a more powerful principle of differentiation rooted in the endless fragmentation and recomposition of labour tasks and production activities in capitalism, and which, in combination with ever-improving technologies of transport and communication, induces continual shifts in locational forces and outcomes. In this process, geography does not become less significant, as some theorists of globalization have asserted; on the contrary, its significance continues to grow. Specifically, the changing structural bases of production and exchange make it possible for firms increasingly to exploit more and more finely grained spatial opportunities for turning a profit. Additionally, as the division of labour proceeds in any sector, streams of externalities and other contingencies are unleashed, often on a massive scale, and with far-reaching locational consequences. Depending on the sector under investigation—as well as on a host of contextual circumstances—these consequences may be expressed anywhere along a continuum ranging from a dominant pattern of spatial agglomeration on the one side to complete dispersal on the other.

The creative field as I conceive it consists of a web of interacting social and economic phenomena at different locations with determinate effects on entrepreneurship and innovation. I use the term 'determinate' here knowing full well that the long tradition of allergic reactions among geographers to any hint of determinism already casts a pall over my meaning. Rest assured, it is no part of my objective here to resuscitate the spectre of geographical determinism in any of its possible guises. That said, I am firmly of the opinion that we need to rescue geographic work from the hallucinating images of free-floating agency that have made such strong incursions into the literature of recent years. My argument is that the different forms of creativity in the modern economy are indeed expressions of free will, but they are determinate in the specific sense that they respond to and work with opportunities that are always concretely situated in a non-subjective world. In the context of the present discussion, these opportunities as it happens are also deeply intertwined with the functional and spatial characteristics of the division of labour.

These two themes come conspicuously together in Chapter 3, which is focused on a specifically geographical approach to the problem of economic development in low- and middle-income countries. In many respects, this problem trumps all the others in modern economic geography. Any economic geography that is worthy of its name must surely be able to say *something* of practical value about what is perhaps the most outstanding human predicament in the world today, and whose urgency can be pinpointed with a single, simple statistical comment. In brief, the poorer half of the world's population today commands just 14 per cent of global GDP, whereas just 15 per cent of the population commands over half of global GDP. Despite the gravity of this problem, Chapter 3 is in several ways the least satisfactory in the book, largely no doubt because of my own intellectual limitations, but also because geographers and economists hitherto have failed signally to come to terms with so many of the most crucial issues in this domain of investigation. There are, of course, outstanding studies by both geographers and economists on development, many of them referred to in the following pages, and I do not want to overstress my complaint. Even so, I think it fair to say that there has been a degree of simple neglect of this problem. Even where research of high quality has been forthcoming it has seemed in recent years to focus above all on macroeconomic theory at the expense of a number of other critical issues, in abstraction from which macroeconomics is just a house of cards. I have tried to push the analysis forward by formulating these issues in terms of what I call *development on the ground*, which is shorthand for the claim that economic development actually proceeds in significant ways via the emergence of urban and regional complexes of productive activity that function concretely as the basic engines of accelerated growth. Above all, I attempt in these pages to draw on the spirit of classical development theory as it was formulated in the 1940s and 1950s, and to show how it takes on new meaning and relevance when recast in terms of modern ideas about the regional foundations of economic activity.

Throughout these three chapters, I am concerned not only to provide meaningful technical analysis of the problems at hand but also to explore some of their wider policy implications. I have therefore been at pains to point out what I think are some of the more fruitful lessons that policy-makers can learn from the discussion. This exercise has compelled me to rehearse once again a number of the basic arguments for and against markets versus policy in the sphere of the economy. The upshot is that I am convinced more than ever that those who argue for a maximum of market organization and a minimum of policy action are deeply mistaken, especially in regard to issues of economic development. Almost any brief encounter with the predicaments of less developed regions and countries should be sufficient, I would think, to persuade any normal individual of the idea that what President Ronald Reagan used to call 'the magic of the market' is vastly overblown. Moreover, not the least of the problems with this highly ideological appeal to the universal efficacy of markets is that for many conscientious individuals it actively undermines their ability or willingness to arrive at a reasonable appreciation of the benefits that markets actually bring to modern society. Surely we can come to a sensible assessment of these benefits while still acknowledging the need for rectification of the widespread social inequalities and irrationalities that are endemic to competitive economies, not to mention the imperative of public regulation of the numerous technical failures of market systems in practice. To balance the books here, I suppose I should remark that there are, of course, many opportunities for policy failure, though it is also pretty clear, given the current ideological onslaught by free marketeers, libertarians, neoliberals, and neoconservatives throughout the advanced capitalist world, that any such warning is superfluous.

I am conscious as I make these remarks that numerous geographers are going to find them unduly focused on economic issues at the expense of many other kinds of social questions, while no doubt a vast majority of economists will feel that they presage yet another series of dogmatic and uninformed attacks on their discipline. To the former my reply is only that while I have been silent on a great many social, cultural, and political issues that intersect with my themes, I have also tried to leave the discussion open on these fronts, though I would balk at some of the extreme reformulations of economic geography that have been proposed under the banner of the so-called cultural turn. To the latter, my appeal must first of all consist of an apologetic admission of my own incapacity to refine my arguments down to their central, logical essentials, but it comprises second of all an invitation to consider the possibility that many useful and important theoretical statements can be made directly and simply in reasonably standard language. Even some of my economist friends (unrepentant quantifiers, at that) complain that they frequently cannot decipher the baroque exercises in mathematical elaboration that pass for being the alpha and omega of respectable discourse in the discipline. For all that, I look forward with anticipation to further research that will enable us to distil the basic issues of geography and economy, as I conceive them, into much more concise analytical language, and that will help us to resolve many of the ambiguities that continue to plague the field. I ought, no doubt, to say something more here about the cultural turn, and add a few remarks about what professional economists increasingly refer to as 'the new geographical economics'. However, I have already tried to come to terms with these different advocacies in a lengthy statement published recently in the *Journal of Economic Geography*, and I refer the interested reader to this piece for further commentary.

In preparation for the writing of these lectures, I have reviewed and re-reviewed an enormous amount of published literature from a wide variety of disciplines. George Eliot says somewhere in *Romola* that 'scholarship is a system of licensed robbery', and the present volume is no exception to this principle. The copious list of references provided at the end of the book bears testimony to this work of pilfering. One of the peculiarities of this literature is that despite its multidisciplinary character it frequently displays symptoms of extraordinary intellectual provincialism, as manifest in the tendency of many authors to acknowledge the research results of only their most immediate circle of academic peers, and only the most recently published results at that. In fact, the central ideas that circulate in this literature have a long and distinguished history, and a great deal of the relevant writing in academic books and journals today picks up on refrains that have their roots in work that was carried out at a much earlier time. I am not referring here simply to such obvious precursors as Smith, Ricardo, Marx, and Marshall, but to dozens of others in the nineteenth and twentieth centuries who have toiled in various ways in this particular vineyard, but now seem largely to be forgotten. A small part of what follows, therefore, entails an attempt to reconstruct something of this lost tradition, and I offer no excuses for the occasional historiographic commentaries and digressions that are scattered through the text, and especially through Chapter 1.

Chapter 2 of this book is an edited version of a paper that originally appeared in *Small Business Economics*. I thank the editors and publisher of this journal for allowing me to reproduce much of the paper here.

In addition, I want to express my gratitude to my hosts in Oxford: Anne Ashby (of Oxford University Press), Gordon

Clark, and Linda McDowell (both of the School of Geography and the Environment). The generous hospitality, friendship, and intellectual camaraderie shown by these and the many other individuals whom I met during my all-too-brief stay ensured that what might well have turned into a series of dry sermons became instead an occasion of genuine human encounter.

CONTENTS

LIST OF TABLES

LIST OF FIGURES

1

Geography and the Division of Labour

1.1 INTRODUCTION

The concept of the division of labour in production has a long genealogy stretching back to the seventeenth century and before, and it recurs repeatedly in the writings of economists and other social theorists down to the present time. In economics, the concept plays a major role in studies of industrial organization, productivity, and trade. In sociology, it has been of major significance as the linchpin of the distinction first proposed by Durkheim (1893) between mechanical and organic solidarity in society. More recently, sociologists have also made considerable use of the concept in studies of the ways in which the division of labour is intertwined with phenomena like race, class, and gender (e.g. Mies 1998; Waldinger and Bozorgmehr 1996). Over the last couple of decades, geographers, too, have made numerous forays into questions of the division of labour and much research has been accomplished on how it ramifies with various kinds of spatial and locational outcomes (Massey 1984; Sayer and Walker 1992). In brief, the concept is of much importance in a wide range of investigations of social structure and dynamics, and it appears to be enjoying something of a renaissance at the present time as social scientists discover or rediscover how profoundly it ramifies with all aspects of modern life.

For geographers, the division of labour has special interest and meaning because, in its role as a mechanism of economic and social differentiation, it is also a fundamental factor in moulding the economic landscape. A peasant society with only weakly developed divisions of labour is not likely to evince much in the way of spatial differentiation except as a function of dissimilarities from place to place in agricultural potentials (themselves related to such variables as soil, climate, and topography). By contrast, economically advanced societies with deep and wide divisions of labour, as in the case of the United States today, exhibit enormous degrees of spatial variation. With the passage of time, moreover, less and less of this variation seems to bear any relationship whatever to underlying conditions of physical geography. In contemporary capitalism, the geography of the world economy as a whole is evidently set on course for eventual reconstruction as an integrated system of differentiated locations based on little more than functional divisions of labour and the ways in which they mould the competitive advantages of different places. Notwithstanding the impediments that stand before the full accomplishment of this ultimate scenario, the history of capitalism hitherto is one in which the division of labour has proceeded, irregularly but definitely, to ever more finely grained locational expression, and to ever wider geographic articulation in a system of production whose limit is in the end nothing less than global. This general process has not meant that the end of geography is in sight, as observers like Cairncross (1997) and O'Brien (1992) have suggested. Rather, the relevance and significance of geography have, if anything, increased. So far from being dissolved away in this process, the economic contrasts between different places have consistently been strengthened by it.

1.2 CLASSICAL POLITICAL ECONOMY AND THE DIVISION OF LABOUR

The modern concept of the division of labour can be traced directly back to Adam Smith (1776/1965). Although Smith did not invent the concept, he was the first to provide an extended and coherent description of the logic governing the fragmentation of work and its relation to market competition. Smith proposes that even in so 'trifling' a case as pin manufacture, a division of labour will tend to materialize, providing that the market is large enough to keep each active worker in full-time employment.

Smith describes the eighteen or so specialized operations carried out in the pin manufactories of his day in terms of such tasks as drawing wire, straightening and cutting the wire, making points, adding heads, tempering, tin-plating, and so on. Each increase in the division of labour brings corresponding increases in output per worker. Smith remarks, in particular, on the vastly superior productivity of the pin manufactory over that of the traditional craftsworker (who makes whole pins from beginning to end). This superior productivity flows from several sources: from the simplification of the tasks to be carried out, from reductions in work set-up times, and from improvements in the capacity of managers to supervise and control the pace of work. Furthermore, extensions of the division of labour tend to result in relative deskilling, thus enabling employers to substitute detail workers for skilled artisans, and thereby to decrease the burden of the wages bill in total production costs. In a more long-run perspective, the division of labour also helps to promote mechanization of production processes, for as it unfolds, it reveals hitherto unsuspected potentials for the substitution of capital for labour in production, including possibilities for general automation of the assembly line. However, at some stage, as Robinson (1931) points out, technological change may result in the *resynthesis* of production processes (with all pin-making tasks, for example, now being subsumed in a single machine), hence establishing an entirely new trajectory of technological and organizational evolution in the sector concerned.

Smith (1776/1965: 17) is at particular pains to declare that 'the division of labour is limited by the extent of the market'. His meaning here is that efficiency in task fragmentation and specialization cannot proceed beyond the point where idle time (Marx's porosity of the working day) begins to cut into profitability. Moreover, by extension of the market, he explicitly means the geographic expansion of producers' market areas. The same argument is made even more emphatically by John Stuart Mill (1848/1909: 130) in the following passage:

The increase of the general riches of the world, when accompanied with freedom of commercial intercourse, improvements in navigation, and inland communication by roads, canals or railways, tends to give increased productiveness to the labour of every nation . . . by enabling each locality to supply with its special products so much larger a market, that a great extension of the division of labour in their production is an ordinary consequence.

The great interest of this statement is that it foreshadows in a number of important respects recent work by economic geographers on the logic of industrial agglomeration. To anticipate a little of the subsequent discussion, a major body of this work pursues the idea that the formation of specialized industrial clusters at particular places is an outcome of agglomeration economies residing partially in the division of labour. As Mill suggests, the division of labour in turn depends to some degree upon the increasing ease with which given clusters can export their outputs to distant parts of the world.

Over the course of the nineteenth century, Smith's ideas on the division of labour were enthusiastically taken up and adapted by many different authors. Among the more noteworthy of these, in addition to Mill, are Babbage (1832) and Ure (1835: 22–3). Almost all later commentators on the issue, however, including even the most partisan, express at least some qualms about the side-effects of an advancing division of labour in society. Ure (1835: 22–3), captures the spirit of this apprehension when he writes about 'that cramping of the faculties, that narrowing of the mind' which is apt to ensue from long daily toil over minutely circumscribed tasks. In the same vein, Jean-Baptiste Say (1803) comments acerbically on the presumed effects of a life dedicated to the production of one-eighteenth of a pin. By Book 5 of *The Wealth of Nations*, Smith himself is expressing misgivings about the effects of the division of labour on the dignity of work, and about the ways in which it promotes forms of employment that are consistent only with pervasive ignorance. Marx, for his part, saw in these aspects of the division of labour some of the most ominous expressions of the oppressive social relations of capitalism. The division of labour may seem to extend the choices available to workers, but in the Marxian view these do not represent real options, only forms of tedious compulsion and human degradation. In *The German Ideology*, Marx and Engels proclaim that socialism will in the fullness of time abolish the division of labour. In socialism, they state in a well-known passage, individuals will finally be able to achieve rounded and non-alienated lives where it will be possible ' ... to do one thing today and another tomorrow, to hunt in the morning, fish in the afternoon, rear cattle in the evening, criticize after dinner, just as I have a mind, without ever becoming hunter, fisherman, shepherd, or critic' (Marx and Engels 1947: 53). For Marx and Engels, then, a key argument in favour of socialism is that it will repair the ravages of the fragmentation of work tasks in capitalism.

The intense interest among classical political economists in the division of labour at the end of the eighteenth century and the beginning of the nineteenth finds further expression in the theory of international trade as laid out by Ricardo (1817). Just as workers in Smith's account come to focus on increasingly narrow labour tasks on the assembly line, so, in Ricardo's analysis, national economies become ever more specialized as trade is opened up between them. The result is what modern theorists would call an international division of labour. Individual countries (like individual workers, at least those who have not yet been reduced to purely *lumpen* status) will tend to specialize in their comparative advantages, reflecting their specific aptitudes and natural endowments. Ricardo proceeds to expound the law of comparative advantage using the example of cloth and wine production in England and Portugal. In Ricardo's example, both of these commodities can be produced more cheaply in Portugal than in England (i.e. Portugal enjoys an absolute advantage in both). However, Portugal's advantage in wine is comparatively stronger than its advantage in cloth, so that if all wine production is concentrated in Portugal and all cloth production in England, total costs will be lower than in the case where autarchy prevails. Portugal's deficit of cloth and England's deficit of wine will then be made up by exchange between the two countries. Ricardo's theory of comparative advantage has been enormously refined by subsequent theorists (cf. Heckscher and Ohlin 1991; Samuelson 1948), and it continues to provide powerful tools for the investigation of international—and inter-regional—divisions of labour. It leads, however, to a rather static view of development and trade and it leaves little or no room for the idea that comparative advantages can be constructed and reconstructed by conscious human effort. Much research on the topic of production and trade today proceeds on the basis of the complementary but considerably more flexible notion of *competitive* advantage as formulated by Porter (1990).

The picture of the world that is presented by much of classical political economy is posited on a view of the sphere of production as an arena of continual subdivision, specialization, and functional reintegration in the context of free markets. It is a picture that acknowledges the human costs of these processes, but, with the exception of a few radical strokes, approves of the broad outcome as an essential condition of economic progress. The picture is further extended towards the end of the nineteenth century with Durkheim's argument about the division of labour as the foundation of organic solidarity in society. As the division of labour proceeds, mechanical forms of social organization (where individuals are merely assembled together under some relation of authority or fealty) give way to deepening ties of interdependence between the different members of society. In conjunction with this shift, the entire moral, legal, and political composition of society at large is transformed; purely ascriptive privileges and obligations start to fade away, and the rule of impersonal law and individual rights is increasingly asserted. Unfortunately, Durkheim did not see fit to extend his analysis of the division of labour and social life to the spatial dimension, though as I indicate at the end of this chapter, it is pregnant with implications for understanding the social morphology of geographic space under capitalist relations of production.

1.3   THE ORGANIZATION OF INDUSTRY

The Division of Labour and Increasing Returns

We must distinguish, at the outset, between two main varieties of the division of labour. One of these involves the fragmentation and specialization of work tasks *within* the individual firm or unit of production. The other involves the fragmentation and specialization of work tasks *between* different firms. In accord with common usage, I shall refer to these two different configurations as the *technical* division of labour and the *social* division of labour, respectively. A very basic set of insights about industrial organization (and eventually about industrial location too) can be put together by tracing out the interdependent logic of these two varieties of the division of labour.

The technical division of labour is further identifiable by reference to Figure 1.1, which illustrates two possible arrangements of work tasks along an assembly line. The upper panel of the figure shows ten different types of task, labeled $a, b, \ldots, j$, each of which is performed by a different worker. In this simple case, the overall efficiency of the line can only be achieved if the tasks to be carried out at each workstation are exactly balanced with one another in terms of workers' inputs of time. The lower panel of Figure 1.1 shows a more complex case where balance is achieved by allocating different numbers of workers to different tasks along the assembly line. The criterion guiding this allocation is minimization of total idle time subject to the achievement of a given overall flow of work (cf. Leijonhufvud 1986). With changes

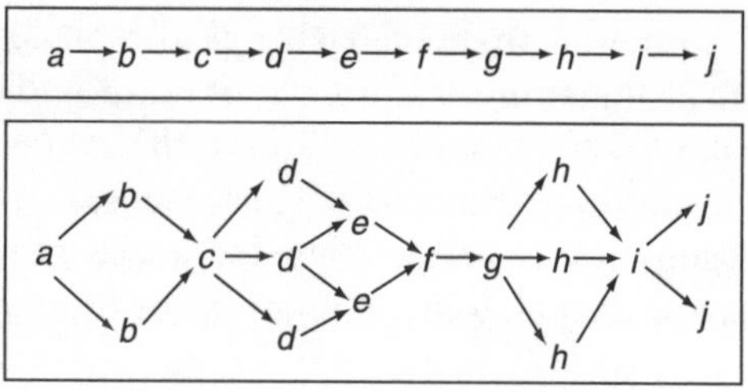

Fig. 1.1. Two possible configurations of the technical division of labour. The terms *a*, *b*, ..., *j*, represent individual workers. The upper panel shows an assembly line where each type of task is carried out by a single worker. The lower panel shows a line where different numbers of workers are assigned to different types of tasks

Source: Adapted from Leijonhufvud (1986).

in the magnitude of this flow, cost minimization proceeds by readjustment over the dual register of the technical division of labour itself and the number of workers allotted to each task. As the flow increases, fine-tuning of the organization of the assembly line can be achieved, leading to enhanced internal economies of scale, that is, reductions in average cost as a result of higher throughput. These operations are not purely mechanical, of course, for they also interpenetrate with the socio-psychic reactions of the workers themselves. Over the last century or so, this particular field of human response has been the scene of many different experiments in industrial relations (Braverman 1974; Friedmann 1956). One main line of experimentation emphasizes the narrow search for worker efficiency, as represented most forcefully by taylorism. A contrasting line is more concerned with worker commitment and responsibility, as represented, for example, by Elton Mayo's human relations approach or modern job enrichment programmes.

Neither Smith nor his disciples in the nineteenth century paid much attention to the possibility that the technical division of labour might disintegrate vertically into a social division of labour. Yet real economic systems in capitalism are always marked by various degrees of vertical disintegration and firm specialization as well as by internal fragmentation of work tasks. Figure 1.2 displays two production systems where the different but inter-related tasks, *a*, *b*, ..., *j*, are now organized in different articulations of the technical and social division of labour. For simplicity in drafting Figure 1.2, I have shown only one firm at each stage (or sector) in the social division of labour. A very preliminary statement as to how any particular articulation of this type comes about can be established by consideration of internal (intra-firm) and external (inter-firm) organizational pressures and their associated forms of governance. Inside the firm, then, product flow between different workstations is organized by managers and imposed by fiat; between firms, product flow is mediated though market mechanisms in response to profitability and price signals. The choice between internalization (vertical integration) or externalization (vertical disintegration) is in large degree a function of the different properties of these two possible states of reality (Coase 1937; Williamson 1975).

One further important point needs to be established before we proceed to more explicit investigation of these matters. The division of labour, in its fully articulated technical and social

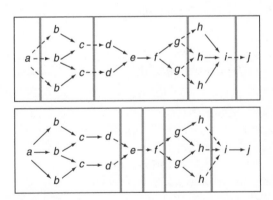

Fig. 1.2. Two possible articulations of the technical and social divisions of labour. The terms *a*, *b*, ..., *j*, represent individual workers. The solid arrows indicate intra-firm flows; the dashed arrows indicate inter-firm flows

dimensions, reflects the number of different but interconnected processing stages that intervene between raw materials and final consumption in any economy. Von Böhm-Bawerk (1891) used the expression 'roundaboutness' to designate this phenomenon, thereby signifying the degree to which the outputs that enter into final consumption in any given economy are dependent on intermediate tiers of production. The same idea was pushed further forward by Young (1928), who wrote that any increase in roundaboutness will tend to enhance economy-wide productivity as a result of the specialization and streamlining of production processes, both within the firm and between firms. Some four decades later, Young's analysis was revived by Kaldor (1970), who re-expressed the relation between the division of labour and productivity in terms of Verdoorn's Law, that is, the principle that the growth of any national economy will promote increasing returns via intensifying roundaboutness. Kaldor went on to propose that Verdoorn's Law applies not only to national economies as a whole but to regional economies as well. The geographic interest of the concept of roundaboutness actually goes well beyond this point of departure, for to repeat an idea that has already been alluded to, *it is also one of the basic conditions making it possible for an economic geography, as such, to come into existence.*

The Coase–Stigler–Williamson Model

In the light of these remarks, the thorny question first posed by Coase (1937) about industrial organization comes insistently to the surface. Why do not all workers function simply as private contractors of their own labour, with every firm being an individual worker and every worker a firm? Alternatively, why is not all production carried out in one gigantic integrated establishment? The organizational patterns that we observe in real production systems are always ranged somewhere on a continuum between these two extreme points, though they always lie closer to the fully disintegrated end than to the fully integrated end.

In setting forth an analysis of these questions, Coase provides an account of how specific configurations of the technical and social divisions of labour are achieved, while simultaneously offering a derivative theory of the firm. The approach adopted by Coase (1937: 395) is to inquire into the processes that establish where the boundaries between firms and markets will be fixed; he writes: 'A firm will tend to expand until the costs of organizing an extra transaction within the firm become equal to the costs of carrying out the same transaction by means of an exchange on the open market'. Given a theoretical starting position equivalent to complete disintegration, then, readjustments in the vertical structure of production will take place, according to Coase, until some sort of equilibrium between the technical and social divisions of labour is reached. The specific pattern of internalized and externalized production activities that eventually emerges is one in which the costs of all transactions within and between firms are minimized. Unfortunately, despite its originality and ingenuity, Coase's account falls short on a number of points of detail. It provides a useful starting point for a composite theory of industrial organization, but it still needs some fairly heavy-handed shoring up on two main fronts. The first of these concerns the cost functions (with their associated economies and diseconomies of scale) that characterize the different stages in the division of labour. The second concerns transactions costs, and more specifically, the explicit role of economies and diseconomies of scope in the organization of industrial systems.

The question of the relations between the vertical organization of production and average costs was initially broached by Stigler (1951), who presented his analysis in terms of two interrelated production activities. Let us designate these activities a and b, where a's output functions as an input to b (as in the case, say, of spinning and weaving in the cotton industry). The average cost functions of these activities are given by c_a and c_b, respectively, as indicated in Figure 1.3. Suppose, in the first instance, that the two activities are vertically integrated in a single firm, with a operating

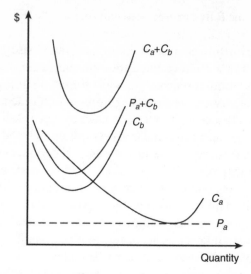

Fig. 1.3. Analysis of production costs for two vertically interrelated activities, *a* and *b*; c_a and c_b are average production cost curves for activities *a* and *b*, respectively; p_a is the price of *a*'s output on the open market

Source: Redrawn from Stigler (1951).

at a rate that is perfectly in balance with *b*'s input needs. Under these assumptions, the firm's total average cost function will be $c_a + c_b$ as shown in Figure 1.3. Now suppose that the two activities are vertically disintegrated. In these circumstances, *b* will secure its inputs from *a* on the open market at the price p_a (corresponding to *a*'s minimum average cost), giving an average final cost curve of $p_a + c_b$. The minimum value of $p_a + c_b$ clearly lies below the minimum of $c_a + c_b$, so that in this example, vertical disintegration is preferable to integration on efficiency grounds. This result, in so far as it goes, illustrates Stigler's principle that where the average costs of the upstream activity, *a*, exhibit increasing returns to scale relative to the average costs of the downstream activity, *b* (and assuming again that both functions always run in mutual balance when they are internalized together in a single firm), then vertical disintegration will tend to ensue. As a matter of fact, the model is more general than this, for there is an incentive to disintegrate no matter whether it is the upstream or downstream activity that displays increasing returns to scale (Scott 1983). It is possible that by running both *a* and *b* at their individually optimal (but unbalanced) levels vertical integration might still be a feasible option, but here we are getting ahead of our story.

Stigler's model takes us part way to an understanding of the problem in hand, but the analysis is still incomplete because it fails to consider the transactions costs that are incurred depending on whether *a* and *b* are vertically integrated or disintegrated. On this matter, the work of Williamson (1975, 1998) is of primary relevance. The point here is that irrespective of the average cost functions of individual tasks, producers also face peculiar costs related to the different ways in which interdependent activities relate to one another within the firm or across markets. Williamson identifies these costs by reference to three specific variables, namely, the frequency, uncertainty, and asset specificity of the transactions between any given set of production processes. Consider the case of *a* and *b* again. Vertical integration of the two processes will tend to occur where their interactions are characterized by (a) high levels of bilateral frequency, (b) uncertainty as to their meaning and content (so that interpretation and decoding are required), and (c) asset specificity, in that the operational specifications of both *a* and *b* are in some sense interdependent. In these circumstances, vertical integration will help to reduce transactions costs and to internalize the benefits of interaction. Where we observe converse characteristics (i.e. low levels of frequency, clarity of information content, absence of asset specificity) a tendency to vertical disintegration will be much more prevalent. Other factors play a role in the determination of patterns of vertical organization in industry, but we shall consider only one further variable at this stage. Thus, vertical integration will also be encouraged where demand for the end-product is stable and predictable so that the firm's investments of capital at different workstations can be optimized relative to the internal flow of materials (Carlton 1979); instability and unpredictability, by contrast, will tend to favour disintegration. It is well to note in passing that transaction costs are also subject to a further variable effect depending on the *distance* between interacting agents.

Obviously, the types of transactions costs identified here assume widely varying combinations of values in different concrete situations, and a full accounting of every possible kind of organizational consequence is far beyond the scope of the present exercise. Let me simply illustrate the main arguments at issue up to this point with the aid of two scenarios, using semiconductor and computer manufacturing as hypothetical examples. Thus, if computer manufacturers have large and persistent demands for semiconductors with subtle qualitative attributes that cannot easily be assessed by anyone but the producer, and where learning effects are likely to be generated by joint management of both forms of production, then there will be a propensity for vertical integration of the two activities to occur, especially if final sales of computers are relatively stable. If computer manufacturers are able to trust the production capabilities of semiconductor makers and if asset-specificity is weak, then vertical disintegration will be the more probable outcome, and it will be even more likely if individual manufacturers' markets are subject to strong and rapid variation. A further possible resolution of the relations between the two activities is conceivable: in certain cases, it may be efficient for an integrated producer of semiconductors and computers to make many more semiconductors than are needed for internal use—notably where the two activities have very different average cost functions relative to final output—and to sell the excess supply on the open market to specialized computer manufacturers. Conversely, if commercialization of the excess supply of semiconductors imposes too great a managerial load upon the firm then disintegration may well prove to be the more economically efficient configuration. In what now follows, I shall frequently refer to benefits that accrue from vertical integration and disintegration as internal and external economies of scope, respectively; and I shall refer to any corresponding costs as internal and external *dis*economies of scope, respectively.

If the insights of Coase, Stigler, and Williamson are combined, a powerful theory of industrial organization can be envisaged, one that is sensitive to the interlaced effects of all the costs and benefits as identified above. In more specific terms, this is a theory that enables us to trace out the articulations of the technical and social divisions of labour in terms of the interplay between economies of scale and scope in both the internal and external dimensions of the production system. A schematic representation of this composite theory is laid out in Figure 1.4. Panel A in Figure 1.4 shows the average cost curve, c_{ab}, for a vertically integrated firm comprising activities a and b, and where it is assumed that c_{ab} is an expression of direct costs plus whatever internal scope effects might be involved in integrated production. Panel B shows individual average cost curves, c_a and c_b, for activities

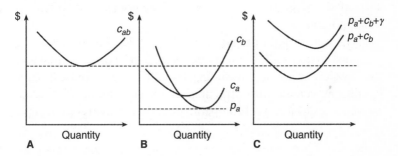

Fig. 1.4. Patterns of vertical integration and disintegration relative to average costs, the market price of inputs, and transactions costs. Panel A shows the average cost curve, c_{ab}, for a vertically integrated firm comprising activities a and b. Panel B shows average cost curves c_a and c_b for activities a and b and the market price of a's output (p_a) under conditions of vertical disintegration. Panel C shows the average cost of vertically disintegrated production, with and without the transactions costs (γ) associated with moving product from a to b. Transactions costs as shown in panel C are assumed to be subject to increasing returns to scale

a and *b* under conditions of vertical disintegration, and where, again, all relevant scope effects have been incorporated directly into the cost functions. The market price of *a*'s output (p_a) is also defined in panel B. Panel C shows the total average cost of vertically disintegrated production, $p_a + c_b$, with and without the spatially dependent transactions cost, γ, that is incurred in moving product from *a* to *b*. The particular significance of Figure 1.4 resides not only in its synthesis of approaches derived from Coase, Stigler, and Williamson, but also in its explicit reference to geographic space, for it points at once towards a model in which transactions costs, in addition to all their other complexities, are a function of distance. Clearly, in the example presented, vertical disintegration without the transactions cost, γ, would be the best possible outcome. When we add γ, however, vertical integration becomes the optimal solution. Only by reducing the magnitude of γ, either by improving transport technologies or by moving *a* and *b* closer together, does vertical disintegration become once more efficient. To anticipate something of the subsequent discussion, vertical disintegration and locational agglomeration often go hand in hand.

As demonstrated in Scott (1983), this style of analysis can be extended to many additional issues of industrial organization and the division of labour, including the efficiency of branch-plant operations versus subcontracting, or the regulation of external transactions by means of competitive or cooperative relationships.

Prototypical Forms of Industrial Organization

We now have a picture of industrial organization processes rooted fundamentally in the notion of articulated technical and social divisions of labour and the manner in which they interweave with the composite play of scale and scope effects and linkage costs. At this point we may fruitfully query the picture for the information that it yields on some of the prototypical industrial systems that recur in capitalist economies. For the present, we shall take it that

Table 1.1. An elementary taxonomy of industrial systems in relation to optimal scale of the representative establishment and external economies

		Optimal scale of representative establishment	
		Low	High
External economies	Low	a. Traditional craftsworker	b. Capital-intensive process industries
	High	c. Disintegrated industrial networks	d. Mass assembly industries

locational variations are held in abeyance, and we shall concentrate our attention only on organizational issues. Table 1.1 lays out the essential points. The horizontal axis of the table represents the optimal scale of the representative establishment in any given industrial system; the vertical axis represents external economies of scale and scope. Ideally, the analysis should distinguish between scale effects on the one side and scope effects on the other, but for expositional purposes, I have selected the option of bundling them together and assuming that they are positively correlated with one another. This is admittedly an oversimplification, but only a small amount of generality is lost as a result. Four basic ideal types of industrial system can now be derived by cross-tabulating the optimal scale of establishment with external economies at contrasting—high and low—levels of resolution (cf. Scott and Storper 1992).

The results of this operation produce the following categories:

(a) The traditional craftsworker, producing outputs like agricultural implements or kitchen utensils on a small scale and operating largely in isolation from other workers.

(b) Capital-intensive process industries (such as petroleum refining or food processing) with very high internal economies of scale, but little in the way of external economies.

(c) Disintegrated industrial networks in which firms tend to be small because internal economies are limited, but where large numbers and a wide diversity of interrelated firms result in abundant external economies. Such networks recur widely in today's new economy, especially in high-technology manufacturing, business services, the media, neo-artisanal production, and the like.

(d) Mass assembly industries, as exemplified above all by traditional car production. In this instance, economies of scale and scope, both internal and external, are usually well developed across multiple tiers of interacting producers.

Of course, hybrids of these four paradigmatic cases are also widely observable in practice. For example, mass assembly industries are typically dominated by large routinized lead plants, but often fade off into more disintegrated network forms of production at their edges (cf. Sheard 1983). In addition, as Storper and Harrison (1991) have shown, composite industrial systems will almost always be associated with overlying structures of governance or regulation, and these structures will vary widely, depending on basic patterns of production and divisions of labour. We would therefore expect to observe rather idiosyncratic institutional infrastructures alongside any concrete instance of the four situations enumerated in Table 1.1.

Our main objective is now to deploy the concepts and analysis developed thus far in a widely ranging investigation of the mainsprings of economic geography. In pursuit of this objective two interrelated sets of issues will be examined. The first deals with the logic and dynamics of intra-regional economic development; the second is focused on the ways in which divisions of labour are expressed at the interregional and international scales.

1.4   THE DIVISION OF LABOUR AND AGGLOMERATION

From Networks to Places

Consider a network of firms caught up in a social division of labour. Any expansion of the market for the final products of this network will be associated with growth and change in at least one of three directions, namely (a) the enlargement of individual firms, (b) an increase in the number of existing firms of any given type, and (c) extension of the social division of labour. The first and second of these outcomes are governed by the structure of intra-firm economies and diseconomies of scale. The third is a more complex kind of event that depends on the interactions between economies of scale and scope in both the internal and external dimensions (recall Figures 1.3 and 1.4). A special but extremely important case for our purposes is represented by the situation where internal economies of scale and scope are limited, but where external economies of scale and scope are relatively strongly in evidence. In these circumstances, our system will tend to evolve as a transactions-intensive network of mainly small and specialized firms. As it does so, the network will generate expanding rounds of external economies. In other words, the network will become a locus of Verdoorn effects, which in turn will translate into system-wide competitive advantages for all.

Within the network, any linkage between any pair of firms will incur a transactions cost whose magnitude will reflect, among other things, the distance over which it is projected. Spatially dependent transactions costs of these sorts are extremely multi-dimensional in practice. They vary as a function of mode of interaction (rail, road, air, etc.), the degree of personal intermediation involved (face-to-face or at-a-distance), the scale of the corresponding linkage (big or small), the substantive content of the linkage (information, perishable materials, bulky objects, etc.), the social character of the interdependencies at issue (traded or untraded, low-trust or high-trust), and so on. These variable attributes bear heavily on the total costs of inter-firm transacting in any industrial system. Above all, costs will tend to be high per unit of activity whenever linkages are (a) small in scale, so that they cannot command cost discounts, (b) irregular over space and time, so that firms are constantly faced with the problems of finding and dealing with new interlocutors, and (c) mediated by face-to-face interaction so that expensive meetings between personnel from different firms are a frequent occurrence. These kinds

of high-cost linkage structures are found persistently in disintegrated industrial networks as identified in Table 1.1, particularly
where these networks are composed of many small-scale producers specializing in non-routine activities so that their input re-
quirements are forever changing, and where inter-firm relations
require personalized intermediation.

In circumstances of this sort, selected groups of interrelated
firms will frequently find it advantageous to locate in close proximity to one another, and they will thus exhibit a propensity to
converge together around their own centre of gravity. This propensity is driven by two sets of circumstances. First, where linkage costs represent a high proportion of overall operating
expenses, proximity is an important means of ensuring that individual firms remain competitive on the price front. Even mass
assembly complexes are subject to similar locational pressures,
most notably in segments that are given over to relatively transactions-intensive modes of operation between adjacent tiers of
producers. Second, the existence of external economies of scale
and scope (both actual and latent) generated by the interaction
and co-presence of interrelated producers will help further to
encourage agglomeration. Alternatively, we might say that producers transform latent scale and scope effects into realizable
agglomeration economies by a locational strategy of clustering.
Inter-firm linkage costs and external economies of scale and
scope therefore frequently work in the same direction as one
another in an intertwined and mutually reinforcing trend to
agglomeration. That said, one possible limit to growth in individual clusters resides in the various diseconomies that are liable
to become more pronounced as a function of overall size. These
diseconomies, however, are never absolute, for in most situations, urban planners and policy-makers work continually to
mitigate the negative effects of emerging barriers to agglomerated
growth and to unleash new rounds of economic expansion.

We should be careful, at this point, to maintain a sharp distinction between inter-firm linkage costs on the one hand, and
external economies of scale and scope on the other. Both sets of

phenomena refer to processes that lie within the external domain of the production system, but the role of each is quite distinct. The cost of any given linkage varies as a function of the bilateral commercial relations between producers, and, as such, its magnitude has no particular relationship to system-wide scale or diversity, at least in a first round of analysis. External economies of scale and scope are determined by the size and diversity of the production system as a whole, and even though in most cases they have some sort of transactional basis—in the sense that their operation involves mechanisms of transmission and receipt— they need to be set apart from inter-firm linkage processes in the narrow sense. Moreover, these external economies flow from a great diversity of localized social and economic activities. Marshall (1890, 1919), identified three specific types of external economy as being the essential components of what he referred to as the 'industrial atmosphere' of particular places. The points of origin of these external economies can be identified under the following general rubrics, though in a terminology that is very different from the one employed by Marshall.

Transactional interdependencies   The co-location in one place of many producers bound together in a social division of labour makes it possible for them continually to fine-tune their production activities by means of frequent readjustment of their input–output relations. This feature is important in disintegrated industrial systems, especially where producers are subject to great uncertainty and market competition, and above all where competition is based on constant product differentiation. The availability nearby of many different specialized types of inputs on short notice also makes it unnecessary for producers to maintain costly stockpiles of materials. In these ways, agglomeration helps to reduce overall levels of risk and to rein in many kinds of operating log jams. Equally, agglomeration helps to lower the costs of circulating capital because reduced inter-firm distances translate directly into reduced product delivery times.

Local labour markets   Many beneficial externalities flow from the local labour markets that develop around clusters of producers (Peck 1996; Storper and Scott 1990). The congregation of numerous workers in one place facilitates the hiring and screening of job candidates by employers, just as the gathering together of widely assorted firms in one place facilitates job-search on the part of workers. The matching of workers and jobs is further facilitated where a variegated supply of different occupations, trades, professions, skills, and so on is available to suit the needs of a diverse mix of employers. The industrial communities within which these processes unfold are important sites of habituation and socialization, in the sense that individuals regularly derive many subtle clues and forms of tacit knowledge about workplace norms and effective job performance from the wider social environment.

The creative field   The competitiveness of dense spatial poles of economic and social activity is reinforced by the creative energies that are unleashed in the daily round of business. Above all, the intense, multifaceted encounters that are endemic within disintegrated industrial agglomerations are the source of endless flows of information, both voluntary and involuntary. The result is a continual circulation of informal, often tacit knowledge about issues of production and work, and the concomitant upwelling of new commercial ideas and insights as complementary pieces of intelligence come unexpectedly into synergistic relationship with one another. As I show in more detail in the next chapter, processes of these sorts are one of the foundations of active entrepreneurship and innovation. Clusters where these processes are strongly in evidence are often referred to as 'learning regions' (Florida 1995).

The external benefits that flow from these three domains of activity are amplified by the play of local hard infrastructures and urban equipment, whose costs per capita are greatly reduced when they are spread out over many users. They are yet further intensified by the emergence of certain kinds of institutional arrangements, such as employers' associations or economic development organizations, notably where these provide coordination services to the local industrial community. Local cultures, too, sometimes function in a manner that boosts agglomeration economies, as in the case, say, where they encourage inter-firm collaboration and cooperation, or where they inculcate particular kinds of work-related sensitivities and capabilities in the labour force.

The Dimensions of Agglomeration

In these ways, the roots of industrial agglomerations penetrate deeply and ultimately into the division of labour in society while being sustained in practice by a great diversity of emergent effects that interact with one another in distinctive patterns of increasing returns and cumulative causation. At the same time, the external economies that flow from agglomerated industrial activities encourage the in-migration of yet more firms and workers, a circumstance that pushes external economies to even higher levels, and so on in round after round of path-dependent growth. Still, these relationships are by no means monolithic in their empirical expression, and many variations in the form and functions of agglomeration can be found.

A reconsideration of the structure of industrial networks can help us to pinpoint some of the more interesting of these variations. We may ask, in particular, what are the possible outcomes to be observed as these networks materialize in geographic space? One possibility, obviously, is that all units of production cluster together in a single large industrial region. Conversely—if linkage costs are low and latent agglomeration economies only weakly in evidence—individual units may be widely scattered across the economic landscape. It is also entirely within the bounds of the possible that any given network will break up into a series of specialized, complementary clusters of different kinds linked to one another in long-distance commodity chains. This result presupposes that transactional densities and increasing returns effects are differentially distributed over the network, with the spatial polarization of specific sub-groups of producers reflecting local peaks of high intensity. Yet another outcome might involve the decomposition of the network into a number of smaller but essentially parallel structures, each of them constituting an individual agglomeration. The latter configuration might be expected to occur where transport and communications costs are so high that they impose tight limits on the spatial extent of markets so that each agglomeration serves a relatively discrete geographic area.

Variations in the form and function of industrial agglomerations are also produced by sector-specific features and prevailing geographic conditions. One very common case, as we have seen, is represented by the classical marshallian industrial district focused on craft industries like clothing, furniture, or jewellery with their internal arrays of small-scale production activities tied together in spatially concentrated networks. Industrial districts of this sort are often embedded within more extensive metropolitan areas, where they are found on many occasions in co-existence with other districts, sometimes sharing with them overlapping labour pools and physical infrastructure. A related case of agglomeration, but at a far broader geographic scale, coincides with large industrial growth centres, some of which extend over whole metropolitan regions, and may even spread out beyond into various hinterland areas. This second type of agglomeration was very characteristic of mass-production systems in the postwar decades, with their large central lead plants and their many-tiered cohorts of dependent input and service suppliers. To be sure, innumerable possible intermediate outcomes or combinations of these two extreme types of spatial agglomeration are conceivable in principle and are frequently observable in practice. These remarks suggest forthwith that any a priori attempt to define industrial agglomerations in terms of their spatial boundaries (except in the loosest possible manner) is not likely to bear much analytical fruit.

With the passage of historical time, overall transport and communications costs per unit of distance typically decline monotonically, and as this occurs, some agglomerations will grow at the expense of others. Krugman (1991) has turned the same point into one of the main pivots of his new geographical economics. Assume for the sake of argument that several functionally identical agglomerations exist, each with its own separate market in geographic space. As transport and communications costs fall, the spatial reach of all producers will expand, and levels of inter-agglomeration competition will intensify. For whatever reason—or even as a purely random event—one agglomeration now moves ahead of all other competing agglomerations, be it ever so slightly. As it does so, its competitive advantages due to localized increasing returns effects will intensify disproportionately, giving it a superior edge in the contestation of markets. If this process continues over a sufficiently long period of time, just one dominant agglomeration of any given type will eventually emerge, while any remaining laggards will stagnate or disappear. David (1985) points to an analogous process that sometimes occurs when competing technologies arise, and that ends finally with lock-in to a single dominant market outcome. For the same reason, an early start down the pathway of development is one of the important factors contributing to the eventual dominance of any particular agglomeration. Even so, the complex dynamics of regional development do not necessarily come to a stop at this point, for the dominant agglomeration itself may be subject to internal technological and organizational changes that shift the internal balance of costs and benefits, leading in one possible scenario to its partial dissolution and reconstitution at more scattered locations. In the theoretical limit when all transactions costs fall to zero and any spillover effects are freed from locational dependency (i.e. when all interactions are carried out by the equivalent of magic carpets) a stage of spatial entropy might be envisaged, though we are quite obviously nowhere near this stage as things currently stand.

Whatever concrete guise it assumes, agglomeration in the sense in which I have described it can also be understood as a sort of proto-urban phenomenon, and one of the driving forces of urbanization in all of its full-blown complexity. This proposition immediately establishes a connection between the division of labour on the one hand, and urbanization as we know it on the other. The internal organization of cities, of course, is deeply affected by numerous social and political activities whose fields of operation are often far removed from the details of the production system and the division of labour, as such. However, there are important ways in which wider urbanization processes repose precisely upon large accretions of capital and labour drawn into one place by the centripetal forces of agglomeration. The intimate interconnection between the division of labour, agglomeration, and urbanization is also surely one of the reasons why modern cities represent such advanced cases of overall functional diversity (in terms of production sectors, job types, occupational strata, worker skills, and associated human attributes) massed together in such narrowly circumscribed geographic areas.

1.5   A HISTORIOGRAPHICAL DIGRESSION

There is now an immense literature on the empirics of the division of labour and spatial agglomeration. I have no intention of attempting to summarize the whole of this literature here (for an extensive review, see Storper 1997). Instead, we will find it useful and informative, I believe, to consider some of the published empirical testimony that was already available well before the round of intensive research activities dating from the 1980s and 1990s moved into full swing. Indeed, the quality of this testimony is in some instances unequalled by latter-day analysts, who by and large tend to neglect this rich vein of antecedent research, apart from routine acknowledgements of the pioneering discussion of industrial districts offered by Marshall (1890, 1919), and an occasional gesture in the direction of Alfred Weber (1909).

The earliest report that I have been able to identify that offers significant empirical information on the workings of the processes under discussion here dates from early in the last century. This is the study of the organizational characteristics of the textile industry carried out by Chapman and Ashton (1914), and it remains in many ways one of the most insightful and informative contributions to the literature. Chapman and Ashton's analysis covers a number of different countries, but their discussion of the British case (in particular, the woollen and worsted industry and the cotton industry) is of special interest and importance.

At the start of the twentieth century, as Chapman and Ashton show, the county of Yorkshire had a total of 76.8 per cent of all the woollen and worsted firms in Britain. Given the manifest status of the Yorkshire agglomeration as the dominant geographic focus of the industry, we would expect it to exhibit relatively well-developed social divisions of labour. And, just so, Chapman and Ashton's data indicate that spinning and weaving activities in the Yorkshire woollen and worsted industry were significantly more specialized than in the rest of the country. Specifically, 64.3 per cent of firms in Yorkshire were vertically disintegrated, compared with only 43.2 per cent of firms in the rest of the country (Table 1.2). Chapman and Ashton provide a parallel analysis of the British cotton textile industry. From its beginnings in the eighteenth century, British cotton manufacturing has been geographically concentrated in Lancashire. In the year 1911, according to Chapman and Ashton, well over 80 per cent of all cotton manufacturing firms in Britain were located in the county, and as many as 81.5 per cent of these were vertically disintegrated into separate spinning and weaving operations. In the rest of the UK, only 48.6 per cent of cotton producers were disintegrated. Chapman and Ashton offer the following comment on the observed functional

Table 1.2. The woollen and worsted industry in the UK at the beginning of the twentieth century

	Yorkshire	Rest of UK	Total
Number of firms:			
Spinning only	245	32	277
Weaving only	332	85	417
Integrated spinning and weaving	321	154	475
All	898	271	1169
Percent of firms:			
Disintegrated	64.3	43.2	59.4
Integrated	35.7	56.8	40.6
All	100.0	100.0	100.0

Source: Chapman and Ashton (1914).

specialization of cotton textile firms in Lancashire: 'The cause is no doubt the economy of business specialization and its possibility in the concentrated Lancashire industry with its convenient marketing centre, its organized commercial system, and its developed means of transportation' (p. 492).

There is a certain vagueness in the quoted passage, but it points in appropriate directions. We can improve upon it by noting that a high proportion of the textile manufacturers in Lancashire at the time Chapman and Ashton were writing were making relatively high quality products for specialized niche markets. Lancashire's most noted specialty was damasked cloth where the pattern is woven directly into the final output, and this feature encouraged manufacturers to introduce frequent variations into final product designs. The many uncertainties and fluctuations attendant upon markets for output like this presumably helps in part to explain the high levels of vertical disintegration observed. Lancashire in 1911 had 657 vertically disintegrated spinning firms and 855 vertically disintegrated weaving firms, the ratio between them being 1:1.3. Thus, specialized spinning operations in the county were relatively large and standardized, with a median of 60,000 spindles per firm, while specialized weaving operations (the segment of the industry closest to final markets) were small, with a median of 450 looms per firm. Vertically integrated firms in Lancashire had significantly fewer spindles (the median being 37,500) and more looms (626), where the given numbers presumably reflect the search for some sort of functional balance within the individual firm.[1] Thus, the final weaving operations of integrated firms absorbed considerably more capital investment than disintegrated weaving did, leading to the speculation that integrated firms concentrated on making relatively standardized final outputs in long runs. Chapman and Ashton do not present equivalent details of cotton manufacturing activities in the rest of the UK. They do indicate, however, that cotton manufacturers outside of Lancashire were significantly more likely to be vertically integrated than disintegrated, and we may surmise also that they were on average at least as large as integrated producers in Lancashire, and possibly more standardized too. Lazonick (1983) has claimed that vertical disintegration became a handicap to entrepreneurial advancement in the Lancashire cotton textile industry after about 1900, but this proposition fails to take account of the dynamics of flexible industrial agglomerations. As Broadberry and Marrison (2002) argue, it is more probable that the external economies of the Lancashire agglomeration actually delayed the ravages that were eventually visited on the industry in the first half of the twentieth century as a consequence of foreign competition.

Additional informative work on the empirics of industrial districts before and including the 1940s can be found in Haig (1927), Allen (1929), Robinson (1931), Hoover (1937), Perrin (1937), Florence (1948), and Wise (1949). All of this work displays shrewd understandings of agglomeration processes, and above all of the propensity for networks of small specialized firms to form distinctive industrial districts or quarters in large cities. The study by Haig (1927), for example, describes the development of the clothing and printing trades in New York in the early part of the twentieth century. Haig's account, which includes a number of meticulously detailed maps, shows how these trades clustered together in Manhattan, forming specialized industrial districts made up of hundreds of small interacting firms. Wise (1949), to cite another exemplary case, traces out the geography of the gun and jewellery trades of Birmingham over the nineteenth century and the first half of the twentieth. Throughout this period, as Wise demonstrates, both trades formed dense industrial districts located close to the centre of the city, each populated by many small firms linked together in deep social divisions of labour. As it happens, the Birmingham gun quarter was greatly affected by a turn to mass production after 1854, and it declined considerably as production was transferred to the large integrated BSA plant at that time, which, symptomatically, was set up at a new suburban location. The jewellery trade, by contrast, continued to thrive until the middle of the twentieth century. Figure 1.5 displays Wise's map of the jewellery quarter in Birmingham as it was in 1948. The map reveals in rather dramatic terms the disaggregation of the jewellery industry into numerous specialized units of production, and their spatial concentration in just one small corner of the city. Pollard (2004) has recently updated Wise's analysis, and she shows that while the number of establishments in Birmingham's jewelry quarter has fallen significantly over the last several decades, its geographic outlines remain more or less identical to what they were in 1948. Wise did not actually relate his study in any explicit way to the division of labour, and in the spirit of the geography of the period in which he wrote, his work is entirely atheoretical, but it is still wonderfully suggestive and pioneering in the light of later investigations. These empirical studies represent the first flowering of a stream of research that was carried further forward after the 1940s by Lampard (1955), Beesley (1955), Hoover and Vernon (1959), Hall (1962), and Jacobs (1961, 1969), to mention only some of the most prominent figures. All of it, in one way or another, deliberates on the formation of distinctive industrial communities and the complex internal dynamics that sustain them. Something of this same tradition can be found in the pioneering research on industrial clusters in the Third Italy that was carried out in the 1970s and 1980s by scholars like Bagnasco (1977), Becattini (1978),

Fig. 1.5. The jewellery quarter of Birmingham in 1948

Source: Reproduced with minor changes from Wise (1949).

and Brusco (1982), at least in so far as their research reaches explicitly and implicitly back to the industrial economics of the 1920s and 1930s.

Over the last two decades, an outpouring of new work all over the world has carried the tradition yet further forward, and has generated an enormous body of additional empirical case studies. As this new round of work started to take off in the mid-1980s, a number of demurrers were registered about its usefulness and generality (cf. Amin and Robins 1990), but the accumulated evidence seems to show that the renewed enthusiasm of the 1980s was, if anything, too restrained. As we now know (see Chapter 3) theories of agglomerated development based on networks of specialized but complementary firms can also be extended well beyond the original geographic frame of reference within which they were first formulated (i.e. the advanced capitalist societies), and have numerous useful applications in less developed parts of the world. We might say that throughout the history of capitalism, agglomerations have been one of the most stubborn and pervasive phenomena of the economic landscape (cf. Pollard 1981), but as I shall now indicate, these phenomena must also be situated within a wider geographic and economic context that ranges over both the national and global scales of contemporary capitalism.

1.6   INTERREGIONAL AND INTERNATIONAL DIMENSIONS OF THE DIVISION OF LABOUR

Up to this point, we have mainly explored the effects of the division of labour on purely local outcomes in what we might call a Smithian–Marshallian world. We turn now to more macro-geographic issues concerning the division of labour at the interregional and international scales. This level of analysis might be characterized as a Ricardian–Listian world, a term that captures both Ricardo's emphasis on locational specialization and trade, and List's arguments (in contrast to Ricardo's advocacy of free markets) about the political mediation of specialization and trade as a means of building up competitive advantage. Elements of a macro-geographic division of labour have always been apparent in capitalism, above all as they came to be expressed in diverse structures of core and periphery interaction over the nineteenth and twentieth centuries.

Core and Periphery at the National Scale

Since the earliest stirrings of the Industrial Revolution, the major capitalist countries have been marked by the emergence of heartland regions that function as powerful locomotives of national growth and development (Pollard 1981). In the eighteenth and nineteenth centuries, these regions sprang into being primarily at locations endowed with coal, iron ore, and other industrial resources, and as they materialized on the landscape they came to form the backbone of the great Manufacturing Belts of North America and Western Europe. The Manufacturing Belt of North America took shape over the nineteenth century with the progressive spread of industry from its original focus in New England and the Atlantic Coast to the Midwest. In Western Europe an equivalent but more geographically fragmented formation could be traced from central Scotland and the British Midlands across Northern France, Belgium, and the Ruhr as far east as Silesia, with outliers in southern Sweden and northwest Italy. Outside these regions, economic activity revolved largely around agriculture, natural resource exploitation, and scattered urban centres providing various kinds of services to surrounding areas.

By the start of the twentieth century this core–periphery pattern of national development was becoming ever more sharply etched on the landscape. The mass-production system was now beginning to move into high gear, and, with steady improvements in its technological and organizational bases, the economic contrasts between core areas and peripheral areas continually widened. The high-water mark of fordist mass production coincided with the period stretching from the 1920s to the 1970s. This was a period when the leading edges of the economies of the advanced capitalist societies of North America and Western Europe were constituted by sectors such as cars, machinery, and domestic equipment. The central operating units of this peculiar type of industrial development consisted of large lead plants, themselves functioning within the multi-establishment corporations that came decisively to dominate business enterprise at this time. Production in these lead plants was organized around extended assembly lines in which technical divisions of labour were often pushed to their extreme limits. Lead plants in turn were the driving engines of the growth poles that Perroux (1961) identified as being the essential axes of capitalist development, and whose voracious demands for inputs set multiple tiers of direct and indirect suppliers into constant motion. In view of the generally high costs of spatial interaction at this time, the lead plants and their associated cohorts of producers tended to form the nuclei of large industrial-urban agglomerations, though this tendency became less assertive in the later period of fordism, which was marked by accelerating rounds of locational decentralization. These industries constituted the economic mainstays of the industrial metropolitan regions that grew so dramatically at this time, and of which Detroit, with its burgeoning car industry, is archetypical.

As fordist industrialization in North America and Western Europe continued to expand apace after the Second World War, it became apparent to contemporary observers that a very distinctive model of regional development and the division of labour was also being played out. Hirschman (1958) and Myrdal (1959), in particular, identified this model in terms of a system of core–periphery interactions. Hirschman couched his ideas in the vocabulary of polarization and trickle-down, while Myrdal used the terms backwash and spread, but there is a remarkable correspondence between their respective positions, as well as in regard to their adoption of cumulative causation as the basic mechanism of regional growth and development. Both argued that large manufacturing centres expand aggressively on the basis of external economies of scale, and that as they do so, they draw in skilled and talented individuals from less well-developed peripheral areas (a process corresponding to polarization or backwash). Peripheral areas, for their part, were described as reservoirs of low-wage labour, and for this very reason were also seen as being attractive locations for deskilled routinized branch plants escaping from high-wage industrial centres in the core (trickle down or spread). One of the great debates that followed from these formulations focused on the relative rates of growth of the core and the periphery. Above all, were incomes rising more rapidly in the core relative to the periphery, or did the converse situation prevail?

Over the entire fordist period, but especially after the Second World War, this core–periphery division of labour steadily tightened its hold on the space-economies of the major capitalist countries, with ever-expanding streams of population moving to the core in search of higher wages and better opportunities, while routinized production units, (above all dependent branch-plant operations) moved to the periphery in search of cheap labour. The standardized operational structures and input–output relations of these branch plants, combined with continually improving technologies of transport and communications, meant that they were becoming steadily less subject to the locational pull of the major centres of production and were hence free to exploit the advantages of low-wage sites. The ground-breaking studies of Creamer (1935) show that in the United States this trend actually began some time before the Second World War, especially in the textile and shoe industries. The radio industry, too, dispersed en masse in the 1930s from its original agglomerated locations in the big cities of the east coast, and regrouped in large mass assembly plants located in small towns in Indiana, Illinois, and Missouri (Lichtenburg 1960). By the late 1940s, as McLaughlin and Robock (1949) indicate, the shift of manufacturing plants from the northeast to cheap labour locations in the American South was affecting a rapidly widening variety of sectors, and the trend continued to deepen over the succeeding decades.[2] Norton and Rees (1979) have documented these events in their study of the rise of the Sunbelt as a major locus of industrial employment over the 1950s and 1960s. At the same time, the postwar decades witnessed an additional steady stream of branch plants from the United States (and Europe) to offshore locations, and eventually this stream began to outstrip purely intra-national locational adjustments between core and peripheral areas.

Up to the mid-1970s, processes of cumulative causation kept the big mass-production agglomerations in the Manufacturing Belts of North America and Western Europe functioning as the dominant hubs of the wider spatial division of labour in advanced capitalism. The decade of the 1970s, however, witnessed a prolonged and debilitating economic crisis of fordism, one of whose symptoms was a yet greater outward rush of production capacity from core areas to peripheral locations, both domestic and offshore. As the crisis deepened the core areas themselves became the scene of widespread industrial dereliction and soaring levels of unemployment, with the result that the US Manufacturing Belt now came more familiarly to be designated as the Rust Belt.

Core and Periphery at the International Scale

Just as the lineaments of a core–periphery division of labour could be discerned at the intra-national scale from a very early stage in the historical development of capitalism, so a parallel pattern could also increasingly be made out at the international scale (Wallerstein 1976). As capitalism expanded during the nineteenth century, the countries of Western Europe and North America became more and more specialized in manufacturing, while huge swaths of territory in other parts of the world were steadily transformed into sources of minerals, agricultural products, and other raw materials destined to serve as industrial inputs and foodstuffs in the more developed economies. The broad outcome was a so-called *old international division of labour* as expressed in a prevalent pattern of world trade in which the industrialized nations exchanged manufactured goods for primary products originating in dependent and colonial territories.

The old international division of labour accounted for a dominant share of world commerce during the nineteenth century, much of it shaped in substantive terms by the geographical distribution of Ricardian endowments, such as mineral and forest resources or agricultural potentials (Yates 1959). Even at this early stage, however, overarching lines of political power also played a role in just how and where these endowments were exploited. The old international division of labour remained a very significant element of the world order until well after the Second World War. It started to recede most dramatically— certainly in relative terms—as colonialism crumbled, and as the politically self-assertive Third World entered the stage in the 1950s and 1960s. Coincidentally with these events, a number of forceful critiques of the theoretical foundations of the old international division of labour began to circulate in selected academic and policy-making circles. Singer (1950) and Prebisch (1959), in particular, mounted persuasive attacks upon the old order by arguing that it worked consistently in favour of the interests of the rich countries and against those of the poor. Both Singer and Prebisch maintained that the terms of international trade between manufactures and primary products were necessarily unequal as a consequence of the contrasting effects of technical progress in the North and the South. In the North, they argued, technical progress was in part translated into higher wages for Northern manufacturing workers; in the South, technical progress was more likely to result only in falling prices of basic commodities, and all the more so given the political disorganization of workers in the South relative to those in the North. Singer and Prebisch contended that even when investors from the North set up manufacturing facilities in the countries of the South, the benefits were largely exported back to the North, with minimal net gains for the South itself. In due course, the analysis set in train by Singer and Prebisch evolved into an elaborate and politically militant theory of dependency and unequal development as represented, for example, in the writings of Amin (1973), Emmanuel (1969), and Frank (1978).

For partisans of the South, an obvious practical deduction from the theoretical work of analysts like Singer and Prebisch was to seek escape from the prevailing international division of labour by means of locally controlled industrialization programmes. In pursuit of this goal, policy-makers in many less developed countries espoused—with varying degrees of success—a two-pronged strategy based on progressive import substitution and growth-pole development in an effort to promote higher levels of national economic autarchy. During the 1950s and 1960s, one important concrete outcome of these policies was the formation of mass-production industrial complexes in a number of countries in the world periphery. Yet despite the theoretical energies and political determination that were invested in import substitution and growth-pole development, these approaches steadily ran out of steam during the 1970s as national markets in peripheral countries became steadily saturated, and as foreign indebtedness rose consequent upon the continued need to import increasingly expensive industrial inputs.

Meanwhile, the long-standing division of labour between core and peripheral countries in world capitalism was apparently evolving into a new configuration consequent upon the great expansion of branch plant operations and international subcontracting activities that was occurring in parts of the world periphery during the 1960s and 1970s. According to Fröbel *et al.* (1980) a *new international division of labour* was now taking shape on the global landscape. The emerging outlines of this new order of things were seen, in a sense, as the ultimate expression of the logic underlying the spatial division of labour in fordist capitalism. Over the postwar decades, the growth and spread of multinational enterprise was creating a situation at the world scale in which, according to Fröbel and his co-authors, the economically advanced countries were coming to specialize more and more in high-order white-collar employment activities such as finance, management, and product design, while many less developed countries were becoming foci of routinized blue-collar operations employing low-wage deskilled labour. At the time of its formulation, the analysis did capture important aspects of the changes then occurring in the world economic landscape (especially where appropriate modifications were introduced to account for intermediate or semi-peripheral areas and so-called newly industrializing countries), but subsequent events have severely eroded its continuing relevance and generality. Core countries have, in numerous instances, greatly expanded their capacity for low-wage, low-skill forms of production, particularly in large metropolitan areas where masses of sweatshop factories can almost always be found. In many cases, the number of these factories is growing rapidly as they feed on ever-widening streams of politically marginalized immigrants from poor countries. At the same time, high-wage, high-skill industrial and service agglomerations have proliferated in many erstwhile peripheral countries over the last couple of decades, often in association with spectacular displays of indigenous entrepreneurial and innovative effort. These developments run counter to the main thrust of the concept of the new international division of labour with its sharply drawn image of a global economy divided between a dominant white-collar core and a subordinate blue-collar periphery.

Elements of the new international division of labour certainly continue to be an important feature of global capitalism today, but the theory that Fröbel *et al.* first put forward over two-and-a-half decades ago must now be seen as explaining rather less of the world than they originally claimed. In any case, the general crisis of world capitalism in the 1970s heralded the rise of a very different model of economic development, one that has brought with it some altogether new directions in the division of labour and the logic of location. The principal components of this model comprise (a) the ascendancy of a 'new economy' (which

I equate here simply with the leading edges of post-fordist economic expansion), (b) the spread of export-oriented industrialization programmes across much of the less developed world (a theme that I deal with in detail in Chapter 3), and (c) intensifying globalization as manifest in the escalating institutional and functional integration of different national capitalisms. We now turn to these issues.

1.7   THE NEW ECONOMY AND THE GLOBAL REGIONAL MOSAIC

Massey (1984) has averred that whenever basic shifts in the structure of capitalist production systems occur, major reorganizations of the spatial division of labour are liable to ensue. The economic transformations that occurred in the aftermath of the crisis of fordism in virtually all the major capitalist societies exemplify this proposition with considerable force. These transformations date above all from the early 1980s, but premonitions of them can be traced back in certain instances to the early 1970s. A shorthand (but far from wholly satisfying) way of identifying some of their essential features is to suggest that they coincide with a historical shift from a predominantly fordist to a predominantly post-fordist regime of accumulation in the world's capitalist economies (Amin 1994).

Whereas the leading edges of the fordist economic order coincided with industries like cars and domestic appliances, the leading edges of the post-fordist or new economy are constituted by activities like high-technology manufacturing, neo-artisanal production, cultural industries, and business and financial services. Clearly, fordist mass-production industries, or at least various neo-fordist modifications of them, have not disappeared from the face of the earth, far from it. But it is also the case that many of the most dynamic foci of production and innovation today coincide with sectors of the new economy whose organizational profiles tend to be relatively de-massified, flexible, and labour-intensive. In comparison with the locational dynamics characteristic of late fordism, the new economy is associated with a definite resurgence of agglomeration processes, in part because of the high levels of vertical disintegration and employment volatility that prevail in many post-fordist sectors. Nor has the large corporation been swept away as these events have come to pass. On the contrary, big multi-establishment and multinational firms are today more common than they have ever been in the past. The major difference is that these firms are now generally much less centralized and hierarchical than the representative fordist-era corporation of the 1950s and 1960s (Dunning 1993). This new style of relatively flat corporate organization is one in which individual operating units typically enjoy a considerable degree of independence from the head office, but are required continually to reaffirm their own viability in terms of basic profitability criteria.

These changes in the underlying regime of accumulation were accompanied by a number of wholly unforeseen shifts in patterns of industrial development and the shape of the economic landscape. Perhaps the most startling of them have been bound up with the multiplication of agglomerations of post-fordist industries at locations far outside the old manufacturing belts. In many cases, thriving agglomerations have sprung up in places whose prospects had been widely considered up to that point as being more or less limited to stagnating traditional sectors or dependent branch plants. The existence of these new industrial spaces became evident in the late 1970s and early 1980s as a series of studies revealed an unusual efflorescence of industrial regions on the basis of craft production in the Third Italy (Becattini 1978; Bagnasco 1977) and high-technology manufacturing in the US Sunbelt (Scott 1986). Something of the same phenomenon was also becoming apparent in the production complexes of East and South-East Asia that were starting to mushroom on the basis of export-oriented industrialization in this period, especially in Hong Kong, Korea, Singapore, and Taiwan. During the 1980s and 1990s, these developmental processes have deepened and widened, leading to a progressive reorganization of much of the contemporary global space-economy, which in important ways can now be described as tending towards a constellation of industrial agglomerations spread out across both more and less developed parts of the world. An essential characteristic of this emerging system is that it is made up of dense localized intra-cluster divisions of labour embedded in far-flung inter-cluster relationships. Individual agglomerations in the constellation are linked together by commodity chains in deepening divisions of labour at the global scale (Gereffi and Korzeniewicz 1994).

As Krugman (1991) has suggested, higher and higher spatial scales of economic integration are quite likely to bring in their train ever larger and more specialized agglomerations, so that the main nodes of this worldwide constellation may well become yet more economically dominant as globalization runs its course. Localized diseconomies may of course make their appearance from time to time and temporarily discourage this growth, though as we have seen these sorts of blockages are unlikely to be permanent or absolute. Perhaps a more telling limitation on the competitive prowess of emerging global superclusters is to be found in the steady shift in the new economy away from simple cost competition, as such, to a much greater emphasis on monopolistic competition *à la* Chamberlin (1933), namely, the differentiation of outputs on the basis of producer-specific or place-specific idiosyncrasies that cannot readily be imitated by competitors. Nowhere is this shift more evident at the present time than in those sectors, from fashion clothing to recorded music and films, whose advantages in domestic and global markets depend upon explicit displays of symbolic content. Even purely utilitarian outputs exhibit mounting symptoms of monopolistic competition, as producers seek to capture specialized niche markets by means of product differentiation. Places that endow local producers with monopolistic benefits of this sort (e.g. as a result of the unique characteristics of the local labour force or the distinctive product designs that flow from established traditions) can remain competitive even where their stocks of agglomeration economies are otherwise quite limited. This means that appropriately advantaged small centres of production will often be able to carve out a special place for themselves in the global division of labour and to ward off, at least to some extent, competitive onslaughts from larger places that produce the same general type of product.

The growth of the new post-fordist economy seems thus currently to be ushering in an economic geography of globalization that is based to a significant degree on an expanding mosaic of interrelated regional economies at various levels of scale and development, though we must not neglect to acknowledge the continuing relevance and power of the intermediate layer represented by the national economy as such. This mosaic is steadily overriding, but has by no means yet entirely supplanted, the pre-existing core–periphery system that prevailed under the old and new international divisions of labour. The principal units of the mosaic, as it currently stands, comprise all major metropolitan areas in the world today (of which there are some 300 with more than 1 million inhabitants), but with an admixture of many, many smaller agglomerations as well (Scott 2001*b*). Commercial relationships between individual agglomerations across the mosaic are deepening and widening at a fast pace, not only in terms of direct import–export and input–output activity, but also in terms of intra-firm trade (itself an expression of the continuing expansion of foreign direct investment) and international production-sharing arrangements. Indeed, intra-firm trade today accounts for as much as half of all world commerce (Buckley and Ghauri 2004). International production sharing is now moving into a new phase of development as entrepreneurs in low-wage countries become increasingly capable of performing work to the specifications of producers in the advanced capitalist countries, and as the relative ease of interpersonal exchange across international borders makes it possible to sustain high-trust relations. One sign of this trend is the steady growth of offshore full-package subcontracting (i.e. the putting out of manufacturing operations in their entirety), as exemplified by the operations of garment producers in Los Angeles and New York whose full-package dealings with firms in Asia and Latin America have grown greatly over the last decade or so (Kessler 1999; Scott 2002*a*). With advancing globalization, then, the individual agglomerations that constitute the mosaic become increasingly integrated with one another in complex relations of competition and collaboration

Some of the more detailed nuances of this story can be highlighted by consideration of the spatial reorganization currently occurring in global audiovisual and media industries. Figure 1.6 represents a schematic attempt to capture some of the main features of this reorganization. Note, first of all, and in contrast to the widely held view, that the geography of the audiovisual and media industries is evolving into something like a unipolar system of production based in the United States in general and Los Angeles in particular. I have tried to represent the industry at some hypothetical time in the not-too-distant future as a polycentric system of agglomerations scattered over the entire globe. It may very well be the case that one dominant centre will continue to outshine the others in commercial terms, but I have presumed that monopolistic competition will make it possible for many other centres, both large and small, to assert their presence on world markets. However—and this is an important point—the ability of these subdominant agglomerations to survive and flourish will also be intimately dependent on their ability to market and distribute their outputs. Figure 1.6 also picks up on a phenomenon that is already plainly evident in the audiovisual and media industries, namely, the tendency for producers in different agglomerations to engage in joint ventures, co-productions, creative partnerships, and so on, with one another. This tendency reflects the contrasting but complementary skills, talents, and capacities to be found in different agglomerations and the synergies that can often be generated when they are combined together. In addition, we may expect to see continuing decentralization of employment from primary centres such as Hollywood, especially in the matter of packages of work tasks that can be disarticulated without undue injury from the rest of the production apparatus and dispatched to relatively low-cost production sites elsewhere. This phenomenon is most clearly manifest in the recent epidemic of runaway production from Hollywood, that is, in the accelerating decentralization of film-shooting activities to satellite locations in Canada, Australia, New Zealand, South Africa, Eastern Europe, and elsewhere (Scott 2005a). Over time, some of these satellite locations may well evolve into viable agglomerations in their own right, either as full-blown centres of motion-picture and television-programme production or as way stations in a globally interconnected system of specialized production locales. Large multinational media corporations have been especially active in utilizing (and sometimes creating) the resources of satellite production centres, but many smaller independent firms, too, are now starting to behave in much the same way. The net result, as

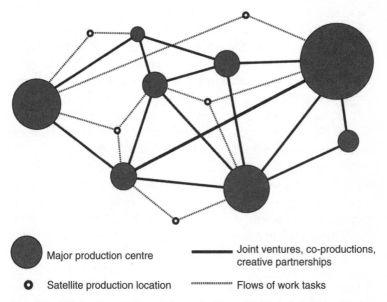

Fig. 1.6. Schematic representation of a hypothesized global production landscape in the audiovisual and media industries

captured in Figure 1.6, is a complex global landscape that combines a Smithian-Marshallian dimension reflecting the dynamics of local economic development, and a Ricardian-Listian dimension, reflecting an overarching structure of economic interaction and place-specific competitive advantage.

1.8   SYNTHESIS: GEOGRAPHY AND THE DIVISION OF LABOUR

I have dwelt at length on the concept of the division of labour, and I have described, first, how it is engendered within the apparatus of production, and then, second, how it becomes visible in multiple ways in patterns of geographic eventuation. However, the division of labour does not function purely as an independent variable with respect to geography, for the latter, in its turn, deeply modulates the specific forms that the division of labour itself assumes. In the context of the accumulative drive of capitalism, these reflexive economic and social relationships in the spheres of production and geographic space lead on to endless processes of disintegration and integration, subdivision and recomposition, and hence also to constantly shifting locational patterns of industry.

In the current globalizing phase of capitalism, marked as it is by a burgeoning international mosaic of regional economies, these patterns are actually becoming more diversified. In short, those accounts of globalization that see it as a process of the dissolution of geographic space into an entropic space of flows have seriously misidentified its underlying dynamics. True enough, globalization is posited upon dramatic reductions in the costs of long-distance transport and communications, but it is also associated with the resurgence and spread of agglomerated economic production and the intensifying spatial differentiation of the world at large. The latter trend is taking place precisely as a function of the ever-growing opportunities for finely grained locational specialization opened up by the falling costs of spatial interaction. In testimony to the same point, Buckley and Ghauri (2004) have shown that multinational firms are now slicing their operations more and more narrowly in both functional and geographical terms in response to the advantages to be obtained from deepening spatial divisions of labour. A noteworthy component of this play of forces concerns those massive agglomerations, or city-regions, that function as the dominant motors of the contemporary global economy. The extraordinary recent upsurge of these agglomerations rests upon a doubly faceted dynamic that seems to be particularly strong at the present time. On the one side, individual producers in diverse industrial sectors today depend for their survival on well-articulated relations of organizational and spatial proximity. On the other side, final outputs made by the same producers often flow with remarkable ease through spatially extended channels of distribution. The first of these conditions is conducive to agglomeration, the second to vast extension of markets, and the interaction between the two is a source of continual and powerful expansion at certain favoured sites. Places like Hollywood, Silicon Valley, and the City of London exemplify the point with some force.

The new world geography that seems to be taking shape as these developments proceed has been evoked here by means of the image of a global mosaic of interconnected agglomerations. As the mosaic expands by the addition of new agglomerations on the extensive margins of world capitalism, it steadily overrides the older core–periphery structure of world economic geography, though elements of this structure may be expected to remain over the long run as a durable subjacent layer of the global economic order. It may well be, too, that at some point in the future, new technological developments and continuing reductions in the spatial costs of transacting will begin to undercut some of the forces at the basis of current trends to large-scale agglomeration (Scott 1998*b*). For the present, the continued intensification and elaboration of the division of labour at many different spatial scales from the local to the global is nothing less than an extension of Durkheim's notion of organic solidarity to world society as a whole. As it becomes manifest in locational form, this phenomenon expresses the enduring and deepening relevance of geography to any understanding of the patterns and processes of contemporary life. In the spirit of Durkheim's original analysis, it also points with some urgency to the need for massive extensions of what he called restitutive law across the many different but interacting communities that constitute the global mosaic, and for their concomitant incorporation into some sort of coordinated arrangement of governance.

2

Geography, Entrepreneurship, and Innovation

2.1 INTRODUCTORY REMARKS

Throughout his voluminous writings, Marx insisted on the notion of capitalism as a turbulent scene of production and exchange, gripped by the forces of competition in an endless process of self-transformation. In these circumstances, every firm faces a stark choice between the continual need to upgrade its process and product configurations or eventually going out of business. The result is what Schumpeter (1942), in an explicit invocation of Marx, called 'creative destruction', that is, the periodic abandonment of old equipment, production methods, and product designs in favour of newer and more economically performative assets. At the same time, as both Marx and Schumpeter recognized, creative destruction is inscribed within an ever-expanding sphere of economic activity due to the growth of existing firms, the extension of entrepreneurship, and the appearance of new products on final markets. Capitalism, in brief, is a complex field of forces spurring constant qualitative and quantitative readjustments across all its multiple dimensions of operation (cf. Baumol 2002). Sometimes these readjustments are of cataclysmic proportions, as when steam replaced water-power in the nineteenth century; more often than not, as Rosenberg (1982) points out, they take the form of small, incremental steps, many of which may be minuscule, but which collectively produce the incessant instability descried by Marx and Schumpeter.

Of late years, there has been a considerable outpouring of literature devoted to these themes, much of it partaking of insti-tutionalist and evolutionary economic theory (e.g. Archibugi *et al.* 1999; Arthur 1990; David 1985; Edquist 1997; Foray and Lundvall 1996; Freeman 1995; Lundvall and Johnson 1994; Nel-son 1993; Von Hippel 1988). An important aspect of this litera-ture is the emphasis that much of it assigns to geography—and above all to the *region*—as an active force in moulding industrial performance *qua* new firm formation, learning, invention, and growth (cf. Acs *et al.* 2002; Antonelli 2003; Audretsch and Feld-man 1996; Cooke and Morgan 1998; Feldman 1994; Howells 1999; Maskell and Malmberg 1999; Oinas and Malecki 1999; Simmie 2003; Storper 1995). This expanding interest in the geographic foundations of industrial performance can no doubt in large degree be ascribed to the emergence of a dominant post-fordist (or more simply 'new') economy since the late 1970s and early 1980s, and to the concomitant transform-ations, often quite radical, of the industrial landscape that have ensued.

For the first three-quarters of the twentieth century, the leading edges of economic expansion in the advanced capitalist societies were constituted mainly by fordist mass-production sectors. Schumpeter himself, or more accurately, the later Schumpeter, identified large firms in sectors like these, with their substantial research budgets and central R&D laboratories, as the principal foci of innovative activity and technical change in the capitalism of that period. Observers of technological change in the postwar decades, such as Mansfield (1968), made much of the distinction between basic and applied research, almost always with the fur-ther observation that the latter was in important ways being pulled along by the former as engineers and other technical workers translated theoretical ideas into practical blueprints for industrial application.

A complementary view of processes of innovation and change in this period of economic history is encapsulated in the so-called product-cycle model (Vernon 1966). Here, the analysis turns on the notion that sectors of production and/or systems of applied technology go through a predictable series of evolutionary changes from their moment of inception to their final expression in the form of mature mass production. The model recognizes three main stages of development in any sector: (a) a period of infancy and experimentation as new technologies and products make their historical appearance and as small entrepreneurial firms spring into existence to exploit them; (b) a period of growth, based on research-intensive process and product development, accompanied by the shakeout of underperforming assets; and (c) a period of maturity or oligopoly in which just a few very large firms making standardized products dominate the entire sector, and in which technological change has radically slowed down. Several attempts were made to incorporate a theory of industrial location into the product-cycle model, as expressed in a composite story to the effect that new industries originate in agglomerated 'incubators' and then steadily disperse outward as they develop, until in the final stages of maturity, virtually all production has decentralized to cheap-labour locations (Norton and Rees 1979; Struyk and James 1975).

In spite of its many oversimplifications and oversights, this vision of technological change and entrepreneurship can be taken as an approximate description of how at least some sectors evolved in the postwar decades. Even in the context of fordist mass-production, however, the product-cycle model fails to provide a really adequate account of technological trajectories and the evolution of the firm (Storper 1985). As the new leading edges of capitalist development today—such as high-technology manufacturing, neo-artisanal industry, business and financial services, the media, and so on—have come to the fore, the deficiencies of the theory have grown yet more apparent, above all in view of the circumstance that one of the defining features of the new economy is its persistent postponement of anything like the stage of maturity. New-economy sectors are endemically given to continuous learning and hyper-innovation in all phases of their growth, not only in so far as tangible technologies are concerned but intangible capital of all kinds as well (Amable *et al.* 1997; David and Foray 2002).

Thanks to the great surge of published research on these matters in recent years, a very much more elaborate and forthright theory of the spatial foundations of creative activity in contemporary capitalism can now be articulated. I have already attempted a preliminary synthesis in this direction in an exploration of the notion of the *creative field* as a critical underpinning of the modern cultural economy (Scott 1999*a*). The present chapter is an attempt to broaden the terms of reference of this earlier work, and to encompass the new economy as a whole, from technology-intensive manufacturing on the one hand to producers of purely symbolic outputs on the other.

2.2   TOWARDS A CONCEPT OF THE CREATIVE FIELD

The notion of a field of creative forces can be used to describe any system of social relationships that shapes or influences human ingenuity and inventiveness and that is the site of concomitant innovations. An adjunct idea is that this field will rarely be frozen in time and space, but that the very innovations it triggers will also act back upon it, thereby causing changes in its organization and operational logic. In the sphere of the economy, such a field might correspond to any number of different organizational arrangements. It might be represented by a system of labour–management relations, a particular type of corporate structure, a certain group of sectors (such as the aerospace industry), or as Leydesdorff and Etzkowitz (1997) have suggested, a 'triple helix' of academic, business, and governmental interests. Freeman (1987) and Nelson (1993) have identified the national economy and its institutional frameworks (or the national innovation system) as yet another kind of creative field.

At the same time, the idea of the creative field goes far beyond specific applications in the domain of the economy. Developments in the spheres of culture and science, too, can in part be understood in terms of arguments that are essentially variations of the notion of the creative field. The social conditions of creativity in art and scientific research have been examined by commentators such as Becker (1982), Crane (1992), Hennion (1981), Livingstone (1995), and White and White (1965), among many others. Authors like these argue that aesthetic and epistemic communities, and the forms of inspiration and inventiveness that they display, all bear mediated relationships to wider social forces and the specific forms of expression that they foster. This is the message of social epistemology more generally, with its emphasis on the essential immanence of all forms of knowledge (Barnes *et al.* 1996; Latour and Woolgar 1979; Mulkay 1972).

In these senses, the broad concept of the creative field has strong affinities with the theory of practice as articulated by Bourdieu (1972) and the structure-agency theory of Giddens (1984). For both of these analysts, human societies consist of a reflexive duality whose basic features entail (a) sets of existing social relations that channel the expectations and behaviour of individual agents in various ways, while (b) individual expectations and behaviour in turn actualize and transform underlying social relations. Neither the relations nor the connections that run reflexively between them and individual agents are hardwired, as it were, but are negotiated out in exceedingly complex processes of human choice and social change. This is not the place to indulge in extended commentary on these theoretical problems. For present purposes, it suffices to observe that in a world that operates in these ways, human practices (e.g. entrepreneurship or innovation) will in certain important respects be explicable in terms of the concrete social relations within which they are embedded, and vice versa, in recursive relationship over time.[1] As a corollary, we can also say that social and economic change in this world will often be wayward but rarely purely adventitious in relation to previous states of order; that is, it will tend to be path-dependent, an issue to which we shall return later.

For the purposes of the present investigation, we will focus our attention on a specifically geographic conception of the creative field in society as a whole. The relevant identification of the creative field for now is that it comprises all those instances of human effort and organization whose *spatial and locational* attributes, at whatever scale they may occur, promote development- and growth-inducing economic change. To narrow the focus yet more, the creative field in this discussion is represented by sets of industrial activities and related social phenomena forming spatially differentiated webs of interaction that mould entrepreneurial and innovative outcomes in various ways. An intrinsic element of this definition is that both the field on the one side and its effects on entrepreneurship and innovation on the other are reflexively intertwined with one another.

This broad idea is in fact far from new, and aspects of it can be found in different formulations in the literature on such topics as the innovative milieu (Aydalot 1986; Camagni 1995; Maillat and Vasserot 1986), the learning region (Florida 1995; Morgan 1997; Storper 1996), regional innovation systems (Cooke and Morgan 1994; Oinas and Malecki 2002), and the like.[2] My present objective is to review this literature in a way that tries to broaden its theoretical bases and that carries some of its hitherto unexamined implications forward onto new terrain. In practice, the over-whelming—though not exclusive—emphasis of the following discussion is on agglomerated economic structures such as industrial districts, regional productive complexes, and urban economic systems. Phenomena like these are almost always characterized by dense networks of firms and multifaceted local labour markets, and as a wealth of published research has shown, these are the settings within which entrepreneurial and innovative energies flourish *par excellence* in the new economy (Acs 2002;

de la Mothe and Paquet 1998; Fischer *et al.* 2001; Hall 1998; Ó hUallacháin 1999; Domanski 2001). Notwithstanding this emphasis on agglomeration, more than passing attention is also paid to much wider spatial frameworks of industrial activity and their increasingly important implications for entrepreneurship and innovation, including, in the limit, the global.

2.3 THE ENTREPRENEUR IN SPATIAL CONTEXT

A fairly common view of the entrepreneur turns on the notion of the risk-taking individual, imbued with animal spirits, in pursuit of self-realization, independence, and prosperity.[3] An allied proposition is that the entrepreneur must also be endowed with remarkable skills and cognitive capacities, especially in the early stages of firm formation when the probability of failure is invariably high (Casson 1982). These notions are often deployed in behavioural investigations of the individual's decision to become an entrepreneur. A strong finding in the literature is that this decision is frequently triggered by some unforeseen contingency, such as the loss of a job due to lay-off or plant closure (Nijkamp 2003).

Another, and not incompatible, view of entrepreneurship is found in the product-cycle and incubator concepts mentioned earlier. Thus, when a new industrial sector emerges (the stage of infancy in product-cycle terms), the pioneering entrepreneurs within the sector are said to depend vitally on certain critical incubation processes. This is a time when the first firms to make their historical appearance normally face highly unstable conditions in regard to technologies, product designs, management practices, and so on. Location in an 'incubator', so the theory goes, helps these firms to survive at a critical stage in their development, and ideal incubators consist of environments offering many different positive externalities (Struyk and James 1975).

These types of environment, it is claimed, occur most often in the core areas of large urban regions with their dense infrastructures, their abundant supplies of rental premises for commercial use, and the diverse services that they offer. Hence, in this view of things, it is in these areas above all that new entrepreneurial ventures will be most likely to flourish.

Neither of these perspectives on the entrepreneur can be said to be plainly wrong, though each leaves much to be desired in terms of analytical penetration and closure, on the one side because the central issues of social context are largely evacuated, and on the other because a rather misleading biological metaphor stands in the way of a more resolute grasp of the social and spatial forces at work. More recent research has greatly improved upon both perspectives by emphasizing notions of socio-spatial embeddedness and by more ruthlessly pursuing the details of the economic logic of entrepreneurship in relation to the dynamics of industrial development. The central hypotheses at work in this new research revolve around the twin notions of networks and social capital (Cooke 2002; Elfring and Hulsink 2003; Noteboom 1999; Westlund and Bolton 2003). Thus, the entrepreneur is not just a lonely individual pursuing a personal vision, but also a social agent situated within a wider structure of economic relationships that can be represented as an actual and latent grid of interactions and opportunities in organizational and geographical space. Any grid of this sort will be composed of more or less densely developed backward, forward, and lateral commercial linkages together with social relationships through which critical information flows continually about business opportunities, resource availability, labour market conditions, and so on. As such, the grid as a whole is also a unit of social capital, that is, a source of benefits to all entrepreneurs or would-be entrepreneurs collectively.

In some accounts, these ideas are further qualified by appeal to the concept of weak and strong ties as formulated by Granovetter (1973). An entrepreneur caught up in a network of strong ties is likely to enjoy high levels of supportive interaction with other individuals belonging to the same network. However, the content of the interaction will be apt to cover a relatively narrow range of information, because strong ties between the individuals of a group lead to constant mutual reinforcement of existing ideas. An entrepreneur with weak ties will receive fainter and less consistent signals, but these will tend to cover a much wider array of information. The ideal network environment for the entrepreneur, or any other type of innovator, as Elfring and Hulsink (2003) point out, is one that involves some balanced mix of strong and weak ties so that individuals on the reception side are likely to pick up an extremely varied mix of stimuli.

So far so good. The network idea facilitates the task of conceptualizing entrepreneurial effort as a socially and spatially embedded phenomenon, but also raises new substantive questions. What is it in particular that defines the order and character of any given network? And how do networks of entrepreneurs evolve over time? Here we need briefly to expatiate again on some of the ideas laid out in Chapter I.

Consider Figure 2.1, which is meant to represent the evolution of a network of inter-firm transactions or linkages and corresponding information flows. I must stress that the figure itself is entirely hypothetical and schematic. It should be viewed only as a simplified abstraction, one possible developmental scenario out of a very large family of alternative scenarios in a path-dependent evolutionary sequence. The initiating event of the changing network structure shown in Figure 2.1 is the establishment of a single new entrepreneurial firm. Let us take it that demand for the type of output made by the firm continues to grow indefinitely, and that as it does so continual expansion in the horizontal and vertical dimensions of the production system occurs. We will assume that as this happens all ventures remain directly and indirectly linked together within a single network. Also, we shall hold technological change for the moment in abeyance. Accordingly, as shown in Figure 2.1, the network will evolve through a series of stylized stages in which each generation is marked by (a) an expansion in the number of establishments

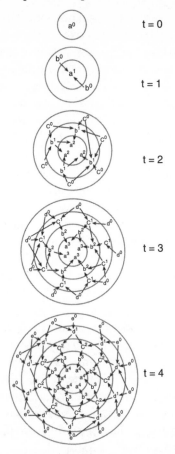

Fig. 2.1. Schematic representation of the vertical and horizontal development of a network of inter-establishment transactions. The symbols a^τ, b^τ, c^τ, d^τ, and e^τ, represent establishments differentiated by sector and evolutionary generation, τ, where τ for any given sector is equated to zero at the time of the first appearance of a new firm or establishment in that sector. The symbol t refers to time in general. It is understood that the changing structure of production from t to $t + 1$ will sometimes be associated with qualitative internal changes in establishments as they evolve from generation τ to generation $\tau + 1$. The arrows represent inter-firm linkages, and these should be interpreted as multi-dimensional sites of product flow, interpersonal contact, and information exchange. We might well expect linkages to occur in the horizontal as well as in the vertical dimension, but this possibility has been avoided simply for the sake of graphical clarity. Linkages will usually be highly variable over space and time

in pre-existing sectors, and (b) the formation of a new vertically disintegrated sector or subsector. This evolving network—irrespective of its locational coordinates—will tend to be a locus of expanding external economies of scale and scope, together with pecuniary externalities *à la* Krugman (1991) and Scitovsky (1954), giving rise to system-wide competitive advantages. It also constitutes a structure or field of entrepreneurial possibilities, meaning that its changing internal order provides a sort of template onto which actual patterns of new firm formation can be (approximately) mapped, providing that there are no deficits of individual entrepreneurial effort and initiative in general. This mapping is expressed in the form of both churning (the refilling of old organizational positions as failed firms are replaced) and developmental change (quantitative and qualitative transformation of the network). In this formulation, entrepreneurship begets entrepreneurship via the latent structure of evolving industrial systems.

As things now stand, Figure 2.1 represents a purely organizational entity lacking any geographical specificity. The firms within the network, whatever its stage of development, may be widely dispersed, strongly clustered, or some combination of the two. Clustered development, however, is highly characteristic of transactions-intensive production systems like the one shown in Figure 2.1, and is especially prone to occur in many of the sectors that constitute the new economy. In more specific terms, as we have seen, producers in such systems frequently have strong incentives to cluster together in geographic space so as to avail themselves of the benefits of mutual proximity in the form of reduced linkage costs and enhanced economies of scale and scope. By the same token, the locational choice of the first entrepreneur, even if it is perfectly random, is liable to turn into a self-fulfilling prophecy by reason of the developmental dynamic that is now set in motion (Krugman 1991). By this I mean that the initial seed that is planted sets off a train of subsequent events in which an organized production network comes gradually into being and is increasingly locked into the initial location by its own expanding stock of agglomeration economies. Moreover, as agglomeration intensifies, individuals already working in establishments within the network are especially well positioned to observe emerging entrepreneurial opportunities as and when they loom onto the horizon, thus pushing development yet further ahead. If this argument is correct, we can expect that a significant proportion of new entrepreneurs in any agglomeration will be established residents of the local area (Romanelli and Schoonhaven 2001).

In line with these remarks, it might be contended that William Shockley's choice of location for his Shockley Semiconductor factory in 1955 in the then backwater of Santa Clara County was random in the sense suggested.[4] Shockley's firm, of course, was the pioneering semiconductor manufacturer in what later came to be known as Silicon Valley. But once Silicon Valley started to develop as an agglomeration of innovative semiconductor producers, success for subsequent entrepreneurs in the same field became increasingly and powerfully keyed in to that same location. This broad analysis, moreover, is consistent with two recurrent sets of observations about entrepreneurial activity in regional contexts. First, firm spin-offs in both the horizontal and vertical dimensions are a commonly observed phenomenon in expanding regional economic systems, as exemplified by the generational trees of firms that have been traced out for the semiconductor industry in Silicon Valley (Assimakopoulos *et al.* 2003) or the medical device industry in Orange County (de Vet and Scott 1992). Second, the empirical evidence suggests that entrepreneurs do indeed have a distinct propensity to establish firms in regions where they already live, and that rates of new firm start-up are especially high in regions endowed with dense agglomerations of producers (Almeida and Kogut 1997; Cooper and Folta 2000; Sorenson and Audia 2000).

Consolidation of any given industrial agglomeration is re-inforced by the formation of what Florida and Kenney (1988) refer to as social structures of innovation, that is, specialized service suppliers, such as venture capitalists, investment bankers, law firms, management and technology consultants, and so on. These suppliers (who are in practice supplementary elements of the evolving production network) help to maintain high local levels of entrepreneurial effort (see also Malecki 1991). Intra-cluster educational and training facilities, too, are increasingly caught up in the developmental process, and often play a major role in supporting new firms. Thus, Zucker *et al.* (1998*b*) have shown, for the case of the Californian biotechnology industry, that entrepreneurs often join forces with leading researchers in nearby university laboratories in order to establish knowledge-intensive start-ups. A further advantage of agglomeration, according to Cooper (2000), is that it facilitates the formation of well-matched founding teams of entrepreneurs; and both Fornahl (2003) and Sorenson (2003) have pointed out that agglomeration enhances each entrepreneur's ability to observe, assess, and learn from the successes and failures of others in the same general field of endeavour.

The argument thus far brings us in some respects full circle back to the notion of the incubator. However, this largely metaphorical idea can now be much improved upon by a more explicit account of the endogenous structure of production networks and the ways in which they help to channel entrepreneurship. These networks have been closely identified here with regional complexes of economic activity, but to repeat, there are many—and probably growing numbers of—empirical cases of production networks that extend over vastly wider geographical ranges, including the national, continental, and global scales, and in which entrepreneurial activities are accordingly equally dispersed. The problem in these instances, of course, is that entrepreneurs who lack the resources for system-wide scanning and the cultivation of long-distance business relations will find it difficult to survive, especially in the early stages of new firm formation.

2.4   SPACE–TIME DYNAMICS OF INNOVATION I: KNOWLEDGE, LEARNING, AND TECHNICAL CHANGE

Frameworks of Practical Knowledge

Practical knowledge is a critical underpinning of all innovative activity. Practical knowledge in turn is structured by its intrinsic relationship to basic science (biology, chemistry, physics, and so on), though the relationship is never one-to-one because cost, demand, competition, and other variables play important roles in how specific pieces of information are actually deployed in the development of new and improved processes and products. Dosi (1982) has proposed that we use the term *technological paradigm* to describe any coherent combination of theoretical science, know-how, and practical applications. He has suggested that paradigms follow evolutionary trajectories in which they unfold according to an internal structural logic, though the process is almost always fitful and incomplete. The evolution of any given paradigm is also a social process to which many different parties contribute via piecemeal accretion of particular fragments of knowledge (Perez 1983). Concomitantly, the innovative potential of any industrial system will be heightened where many of these fragments are sufficiently different from one another, yet intra-paradigmatically related, so that their combination yields new practical insights (Antonelli 2003). This is a slightly different way of expressing the idea that a balanced mix of weak and strong links is likely to be more synergistic than a set of purely strong or purely weak links. Noteboom (1999) has made essentially the same point in his account of the impact of interpersonal cognitive distance (further refined in terms of novelty and communicability) on learning. Too little novelty or communicability is unhelpful, and so is too much; intermediate doses of both are calculated to push learning forward most rapidly.

Technological paradigms, then, evolve over time by means of internal incremental changes generated by different participants in the knowledge-production process. Occasionally a paradigm will become exhausted and will cease to yield much in the way of new insights and competitive advantages. On other occasions a paradigm shift may occur as a superior new technology emerges, resulting in a rupture in prevailing patterns of industrial development. The same rupture will sometimes be accompanied by radical readjustments in the locational structure of production, with corresponding changes in the geography of entrepreneurship and innovation. The point is dramatically exemplified by the crisis of the Manufacturing Belt and the growth of the Sunbelt after the early 1970s as post-fordist industrialism started its spectacular ascent (Storper and Walker 1989). Steed (1971) has documented a more sharply focused case of technological-locational rupture in his study of the demise of traditional linen manufacturing in Ulster after the Second World War in the face of competition from the international synthetic fibre industry.

The ideational core of any technological paradigm is composed of what can be identified in the terms proposed by Polanyi (1966) as both tacit and explicit knowledge (Foray and Steinmuller 2003; Gertler 2003; Lissoni 2001; Pinch *et al*. 2003). Tacit knowledge is describable as a kind of inarticulate sense of how things work, which is embodied in particular individuals in particular organizational settings. Tacit knowledge is difficult to transmit to others and often can only be transferred by means of close personal interaction and practical demonstrations. By contrast, explicit knowledge is codified or codifiable in ways that make interpersonal communication relatively easy. Polanyi himself illustrates the difference between the tacit and the explicit by reference on one side to the know-how involved in being able to drive, and on the other to the systematic theory of the internal combustion engine. This illustration underlines Polanyi's contention that our total personal stock of knowledge is almost always more extensive than our ability to inscribe it all in formal texts. Since tacit knowledge is specific to given individuals, and may also be culturally encoded, it is especially difficult to transmit over long distances, whereas explicit knowledge can be more directly and cheaply transferred by means of formal inscription (Antonelli and Calderini 1999; Gertler 2003), though Foray and Steinmuller (2003) argue that new ways of recording tacit knowledge (with the aid of digital technologies) are likely to enhance its transferability to wider circles of recipients. In practice, the distinction between the tacit and the explicit is rarely cut and dried, and most forms of knowledge are a complex amalgam of the two types.

Stocks and Flows of Knowledge in the Creative Field

Whatever its paradigmatic features, or its tacit/explicit dimensions, economically useful knowledge is always unevenly spread out over many different sites at many different geographic scales from the local to the global (Archibugi *et al.* 1999; Bunnell and Coe 2001). Typical active sites of knowledge accumulation in the creative field are the individual worker, the firm, or the sector; other sites are represented by specialized institutions such as schools and universities, research laboratories, labour unions, or trade associations. Such sites constitute the atoms and the neurons of the creative field, so to speak, but their power to generate new knowledge is magnified many times over when they come into definite interrelationship with one another. A central question at this stage is what kinds of knowledge accumulate at different sites, how are the bits and pieces interconnected, and how do the concomitant flows and interchanges promote spatially determinate forms of process and product innovation?

The pressures of competition in capitalism make it imperative for firms continually to revitalize their core competencies in the search for production and marketing advantages. The knowledge that enables them to do so comes from two main sources. First, firms acquire knowledge by dint of learning based purely on their own internal resources. Learning-by-doing is one of the most common means by which they do so, above all in the case of small firms, which usually do not have the wherewithal to engage in formal research (Antonelli and Calderini 1999). Large firms, by contrast, are frequently capable of carrying out in-house R&D, and the results of this activity play an important role in identifying breakthrough innovations. Second, firms also learn by appropriating knowledge produced by external sources, such as other firms, or institutions like universities and government laboratories. The pathways by which knowledge spills over in this way are many and various; they include written texts, informal conversations, input–output links, inter-firm mobility of workers, strategic alliances, and so on. In this manner, knowledge produced at one site is acquired, often *gratis*, at other sites, but almost always in ways that reflect some underlying spatial bias (Anselin *et al.* 1997; Audretsch 2002; Feller 2002; Grossetti 1995; Maskell and Törnquist 2003; Rodríguez-Pose and Refolo 2003; Varga 2000; Zucker *et al.* 1998*b*).

We need to make a fundamental distinction here between innovations that flow from aprioristic R&D programmes, and innovations that occur, as it were, by coincidence. In the former case, we know what we are looking for (e.g. supercomputers, fuel cells for electric vehicles, or biological cloning mechanisms), even though we may not have a very clear idea about how to arrive at the desired destination. In the latter case, we stumble across better ways of doing things or better product configurations, more by accident than by design. An especially important form of innovation of the latter type occurs when firms engage in complex business transactions with one another, especially where these entail a good deal of preliminary discussion and mutual assessment. Individuals caught up in these discussions will often arrive at insights that would otherwise have remained hidden from them. For example, Russo (1985) shows how small-scale, informal, but cumulatively significant innovation occurs in the tile-producing district of Sassuolo in the Third Italy, as manufacturers and their suppliers of machinery negotiate with one another about the specifications of new orders for equipment.

In a somewhat similar vein, Von Hippel (1988) has pointed to the important role of information feedback from users to manufacturers of surgical instruments. Indeed, manufacturers typically rely on such feedback as a major source of ideas for product improvement and innovation (see also Lundvall 1988; Lundvall and Johnson 1994). In some instances, groups of firms build specific managerial mechanisms such as joint ventures or R&D alliances in order to enhance the exchange and accumulation of knowledge, with the biotechnology industry being an outstanding example of this propensity (Powell *et al.* 1996). By all accounts, inter-firm flows of knowledge are a pervasive phenomenon in the world of contemporary industry, and are critical stimuli to innovation. A sample of relevant studies on this matter might include, for example, Cumbers *et al.* (2003), Edquist (1997), Gertler (1995), Powell *et al.* (1996), Rallet and Torre (1999), Uzzi and Lancaster (2003). Most of these studies put a heavy, but by no means exclusive, emphasis on the importance of locational proximity as a prime requisite for the successful transmission of knowledge between different parties. In cases where face-to-face intermediation of tacit knowledge is at stake, the role of proximity is especially critical. A basic point that now needs to be pressed home is that this spatial condition reaffirms the major role of industrial agglomeration in the articulation of the creative field, especially in the case of new-economy sectors with their transactions-intensive structures of interaction (Audretsch 2002; Morgan 2004; Scott 2000; Simmie 2003).

Empirical confirmation of the powers of spatial agglomeration in regard to knowledge-generation can be found in the empirical work of Acs *et al.* (2002), Jaffe *et al.* (1993), Ó hUallacháin (1999), and others on the geography of patenting. This work suggests that patents originate with a high degree of likelihood from agglomerated centres of production, as opposed to places where industrial production is less densely developed. A parallel line of research based on data compiled by the US Small Business Administration underscores these results by consistently pointing to the spatial concentration of innovative events in industry (Acs

2002; Audretsch and Feldman 1996; Feldman and Audretsch 1999; Feldman and Florida 1994). The innovative activities of small firms appear to be especially susceptible to stimuli originating in agglomeration processes (Almeida and Kogut 1997). The work of Jaffe *et al.* (1993) is notably significant as a confirmation of the positive impacts of proximity and agglomeration in the creative field because it actually traces out direct lines of influence from one patent to another as revealed by the citations to prior patents that accompany any application to the US Patent and Trademark Office for patent protection. On the basis of a large body of data on this phenomenon, Jaffe *et al.* show that cited patents originate with high levels of probability from the same geographic locality (state and metropolitan area) as citing patents.

Once this has been said, patents are notably troublesome as a measure of innovation, because not all innovations are patented, and not all patents are equally innovative or rewarding. These and other ambiguities of patent data as a measure of innovation have often been pointed out (see, in particular, Griliches 1990). For this reason, the empirical work cited in the previous paragraph is far from conclusive, even if it tends to point fairly consistently in one direction. A particular problem with much of this work is that it proceeds largely on the basis of aggregate measures of patenting activity or innovation, in which data for many different sectors are pooled together. Disaggregated analysis would presumably indicate that great variety exists from sector to sector in regard to rates of patenting/innovation as a function of locational clustering, with some sectors exhibiting considerable responsiveness while others remain more or less unaffected. Moreover, whereas a high proportion of published studies based on aggregate data reveal that there is a statistically significant and positive relation between patenting and/or innovation and the clustering of firms, they rarely display evidence of the increasing returns to cluster size that we would expect on theoretical grounds. Beaudry and Breschi (2003), Breschi (1999), and Lamoreaux and Sokoloff (2000) show that numerous cases exist of industrial clusters that fail to demonstrate any proclivity whatever to innovation. Clearly, more refined analyses are required to push our understanding forward here.

One further line of research on the effects of agglomeration on industrial innovation and productivity merits attention in this context. I am referring here to econometric work that proceeds on the basis of a fundamental division of agglomeration economies into so-called localization economies (or Marshall–Arrow–Romer externalities) and urbanization economies (or Jacobs externalities) (Baptista and Swann 1998; Capello 2002; Feldman and Audretsch 1999; Glaeser *et al.* 1992). The former represent externalities produced and consumed only in a given sector; the latter are defined as externalities that flow between firms in all sectors. In some studies, localization economies are found to be dominant (e.g. Baptista and Swann 1998); in others, urbanization economies are more prevalent (e.g. Feldman and Audretsch 1999); in yet other cases, both types appear to be at work (e.g. Capello 2002). On due reflection, this ambiguity is scarcely surprising. The simple reason is that while these measures obviously pick up on certain kinds of industrial responses to agglomeration, they are incoherent substitutes for other more directly relevant variables, for from what has gone before—not to mention the wider literature on industrial districts—we may infer that the critical issue is less the origin of externalities in this or that sector or group of sectors, than their roots in learning processes (together with transactional networks, local labour market structures, infrastructural artefacts, and so on). When econometricians claim to find evidence for localization economies or urbanization economies, they are not uncovering fundamental dimensions of innovation, but only dark mirrors through which more basic processes are being reflected.

Cultural and Spatial Differentiation of the Creative Field

Spatial relations of proximity and separation exert profound effects on the functioning of the creative field, but cultural variations between different social groups and different places also modify these effects in very tangible ways. A shared culture is often a significant asset in promoting knowledge exchange and innovative effort, just as cultural differences can result in costly misunderstandings, particularly where tacit knowledge is involved. Gertler (1995) has documented a number of disruptive misunderstandings between Canadian users of advanced process machinery and German producers, due to culturally distorted flows of information and contrasting codes of reference. In a complementary vein, Nonaka (1994) has emphasized that common patterns of socialization promote more effective communication of tacit knowledge. In circumstances where bonds of trust have been established, communication is likely to be even further enhanced (Cooke and Morgan 1998; Cooke 2002; Uzzi and Lancaster 2003). A vivid illustration of the play of cultural factors in processes of communication and innovation is presented in the now classic work of Saxenian (1994), which traces out contrasts in the changing fortunes of high-technology firms in Silicon Valley and along Route 128 over the 1970s and 1980s. Firms in Silicon Valley were found to be relatively open to interchange with one another, whereas firms along Route 128 developed inward-looking cultures that effectively insulated them from incursions of new ideas. As a result, according to Saxenian, firms in Silicon Valley displayed a distinctively greater propensity for adaptation, innovation, and survival than those located along Route 128.

Once again, the important role of agglomeration as a nexus of performative intensity is underlined in these remarks. Localized clusters of firms and workers are sites of intense and recurrent daily interaction, and they are, by the same token, scenes of at least some forms of common socialization and cultural development. Local residents often acquire shared understandings and codes that ease interpersonal communication and that facilitate the formation of fresh insights in the workplace (Breschi and Malerba 2001; Brown and Duguid 2000*a*). Certain groups of residents and workers may form tightly wrought technical and epistemic networks, or what Wenger (1998) calls communities of practice, endowed with unique kinds of problem-solving capabilities. Still, and despite the apparently uplifting case of Silicon Valley as described by Brown and Duguid (2000*a*) and Saxenian (1994), we should not exaggerate the tendency for community-wide cultural norms to function as a source of positive synergies. Examples of lock-in to more problematical outcomes abound in reality, including the cases of traditional cultures that are in various ways at odds with the functional essentials of capitalistic development and growth.

Notwithstanding the emphasis on agglomeration in much of the above, we must recall that the geography of industrial innovation also needs to be set within a much more extensive spatial context. The nation is a critical nexus of social forces constituting a distinctive innovation system. Increasingly, too, much innovative activity today resides in relationships that are nothing less than global in extent (Amin and Cohendet 2004; Simmie 2004). Multinational corporations are a major factor here. Moreover, whereas R&D activities were once thought to be congenitally tied to domestic locations (see e.g. Pavitt and Patel 1991), the evidence now indicates that multinationals are increasingly prone to spread their research and scanning activities across multiple international sites (Cantwell and Janne 1999; Dunning 1993). A noteworthy detail here is that the foreign R&D laboratories of multinationals have a special affinity for locations in specialized agglomerations where they can tap into and appropriate local expertise and then re-diffuse it through their intra-corporate networks (Cantwell and Iamarino 2002; Chacar and Lieberman 2003; Cohendet *et al.* 1999; Ernst and Kim 2002). This re-diffusion is accomplished by long-distance transmission of information complemented by occasional face-to-face meetings of key personnel. In fact, it is common practice in both the corporate and non-corporate worlds to enrich the flow of information between individuals normally located far from one another, by means of temporary gatherings (quasi-agglomerations) such as conferences, seminars, workshops, consultative meetings, and so on. These gatherings present an opportunity for brief but intensive inter-communication in highly personalized situations, after which the participants disband back to their scattered work sites where they continue to interact on the basis of the knowledge and insights acquired during their face-to-face encounters.

New communications technologies are now bringing about major shifts in information-flow processes. Not only is it becoming possible to transmit ever larger quantities of explicit knowledge over greater distances at decreasing cost, but much tacit knowledge as well. Some analysts, such as Kaufmann *et al.* (2003) and Leamer and Storper (2001), have claimed—not incorrectly—that the capacities of the Internet are limited in this respect because it does not easily lend itself to ostensive interactions. However, if the speculations of Cohendet *et al.* (1999) and Foray and Steinmuller (2003) turn out to be on track, we can expect considerable relaxation of this limitation to occur in the future, as the Internet, in combination with embedded workstations, becomes increasingly capable of handling information transfers of enormous complexity and subtlety.

2.5 SPACE–TIME DYNAMICS OF INNOVATION II: CULTURE, SENSIBILITY, AND SYMBOLIC PRODUCTS

One of the striking features of the types of knowledge, learning, and innovation that we have considered thus far is that they tend to be cumulative: one discovery leads potentially on to another in round after round of evolutionary progress. By contrast, there are important facets of the modern economy that are given chronically to the search for novelty, but that do not display much cumulative development in the guise of better and more efficient ways of doing things. Fashion-intensive industries, such as clothing or jewellery, are obviously strongly subject to this syndrome, as are the media and entertainment sectors. More generally, cultural-products industries as a whole, as represented by motion pictures, music, electronic games, architecture, or tourism, and the fashion industries generally, are all engaged much of the time in the pursuit of novel but essentially non-incremental variations in output configurations. These cultural-products sectors represent a major and growing share of employment and output in the new economy (Power 2002; Pratt 1997; Santagata 2004; Scott 2000), and they are the core elements of a rapidly widening system of symbolic production in contemporary capitalism. To be sure, individual industries in the cultural economy are also susceptible to radical technological change (Schweizer 2003), but the main emphasis in the present discussion will be on the representational and stylistic dynamics of their final products.

Just as conventional manufacturing industries fall under the sway of technological paradigms and trajectories, so cultural-products industries are subject to the play of design archetypes, that is, basic frames of coded references within which elements of symbolic content and style can be endlessly combined and recombined. Given archetypes are in one sense free-floating phenomena, in that they can be widely imitated, but they are also often intrinsically associated with a particular point of origin, such as a firm, region, sector, or nation. Like technological paradigms, design archetypes are subject to radical structural shifts, sometimes because of variations in final market demand, sometimes because of technological or organizational change in underlying production processes. The history of Hollywood over the last century provides a number of illustrations of the latter type of change, most notably the aesthetic transformation of dramatic content that occurred as the classical studio system gave way to the so-called New Hollywood after the 1940s and 1950s, and then, again, as special-effects technologies matured in the 1990s (cf. Bordwell *et al.* 1985; Scott 2005*a*).

Cultural-products industries are in general exposed to high levels of uncertainty and risk due to the combined effects of unremitting product differentiation on the supply side and the fickleness of tastes on the demand side, even where these effects do not entail basic changes in design archetype. In some sectors (such as music) the instabilities are compounded by the emergence of what Peterson and Berger (1975) have called 'cycles of symbol production' in which large firms (or majors) and small independent producers vie with one another in rotating sequence for market share[5] (see also Hirsch 1972). Because of this volatility, and the ways in which it translates back through the production apparatus of the firm, cultural-products industries are especially given to vertical and horizontal disintegration (Caves 2000; Scott 2002*b*; Storper and Christopherson 1987). Production is thus very often spread out over networks of many different firms. In addition, numerous sectors in the cultural economy exhibit strong signs of industrial dualism as represented by dominating groups of majors complemented by masses of smaller independents. The dense interlinkages that run vertically and horizontally through these sectors means that cultural-products industries generally have an inclination to form dense agglomerations, and this tendency is reinforced by their massive aggregate labour demands and the external economies that flow from the co-presence of many interrelated firms in one place. Such agglomerations occur especially in large metropolitan areas like New York, Los Angeles, Paris, London, or Tokyo, where they form distinctive industrial quarters or districts.

The transactions-intensive structure of the cultural economy is particuarly characteristic of sectors like motion pictures or music recording, in that production is frequently organized around specific projects in which different firms and freelance workers coalesce functionally together, only to part company again when any project is completed, and then to re-coalesce in new combinations as further projects come along. Such interaction, as we know from the previous discussion, promotes the circulation of ideas and is a stimulus to creativity. Additionally, many firms, especially those that put a premium on imaginative product designs, organize their internal operations around temporary project-oriented work teams in which regular employees, part-time staff, and freelance specialists combine together to pool their expertise and talent (Bielby and Bielby 1999; Blair, Grey, and

Randle 2001; Grabher 2002; Sydow and Staber 2002). Shifting, open-ended teams of this sort are often capable of multiplying the creative powers of their individual members many times over (Nonaka 1994). Arresting examples of this phenomenon are offered by Hennion (1981) and Kealy (1979) in their accounts of how popular music recording sessions proceed through sequential adjustment as different participants, from the performers to the recording engineers, respond to one another's comments and suggestions in actual working sessions. Grabher (2001) has described the organization of work in the advertising industry in somewhat analogous terms.

This multiple and constantly shifting transactional structure in cultural-products industries means that much of the workforce becomes enmeshed in a network of mutually dependent and socially coordinated career paths (Montgomery and Robinson 1993). If anything, this condition is even more pronounced in the cultural economy than it is in other sectors of the new economy, and it is a powerful mechanism of general socialization and habituation. Equally, it reinforces the effects of other mechanisms tending to encourage the formation of distinctive cultural communities in particular places. Workers in these communities thus not only develop complementary technical skills, but also come to share sensibilities and mental attitudes that help to boost their joint creative capacities within given design paradigms to yet higher levels. A workforce moulded in these ways is an exceptionally valuable asset in production systems where transfers of tacit knowledge are a key element of the labour process; and it is of particular moment in the cultural economy where competitive prowess derives above all from the distinctive aesthetic and semiotic content injected into final products. In addition, densely developed industrial agglomerations are almost always endowed with formal and informal workers' associations that bring workers into even closer mutual interaction. Cultural-products agglomerations are rife with such associations, which are an important means of reducing the abnormally high-levels of uncertainty and risk to which creative workers are typically exposed (Benner 2003). Workers participate in these associations both to facilitate the acquisition of labour market information and to keep abreast of new developments in their specialized fields of activity (Scott 1998*a*).

Communities of cultural-products workers and their associated production systems are the *loci classici* of what Florida (2002) has called the 'creative class', though Florida's definition of this social category includes workers in a far wider group of sectors than cultural-products industries as such. Ambitious and talented individuals in search of professional and personal fulfilment find these communities irresistible, and they accordingly flock in from every distant corner in a process that Menger (1993) has referred to as 'artistic gravitation'. As a consequence, the labour pools of dynamic cultural-products agglomerations are continually subject to replenishment by selective in-migration of workers who are already predisposed to high levels of job performance even in advance of their arrival.

An additional ingredient in this rich creative mix of production networks and local labour markets is place itself, not only as a collection of industrial capabilities and skills, but also as a stockpile of traditions, memories, and images that function as sources of inspiration for designers and craftsworkers, and that help to stamp final products with a unique aura (Drake 2003; Rantisi 2004). Thus, Parisian fashions, London theatre, Nashville music, or Scotch whisky, are not just generic fashions, theatre, music, or whisky but authentic expressions of an accumulation of past accomplishments. They accordingly acquire distinctive reputations, which means in turn that they can be imitated but never perfectly replicated elsewhere (Molotch 1996). This association between place and product, moreover, is self-reinforcing because both of them are joined together through the interdependent perceptions that consumers have of them. Old established images are regularly modified and recycled through the production system, while new ones continually contribute to the enlargement of local repertoires of place-specific symbologies and designs, in round after round of reciprocal enrichment.

By the same token, the capabilities and reputations of the individual firms that make cultural products and the places where they are located almost always imbue final outputs with unique competitive advantages on consumer markets. In the terms proposed by Chamberlin (1933), markets for such products are endemically given to monopolistic competition. This condition manifests itself in part in the addiction of cultural-products industries to the quest for novel product configurations. But so, too, is it evident in the occasional cases of firms that resist notable change in product designs over long periods of time because of their unique reputations on consumer markets. Thus, once a firm has established a brand with a durable reputation for quality (Louis Vuitton handbags, or Wedgwood pottery, or Rolls-Royce cars, for example), it has a strong incentive to maintain the basic shape and form of its products. Even in these cases, periodic fine-tuning of final designs is prone to occur in response to market shifts.

The rapid rise of cultural-products agglomerations in almost all of the advanced capitalist societies in recent years goes hand in hand with cognate transformations in the wider urban environments in which they are ensconced. Cities in which high proportions of the labour force work in cultural-products sectors often express this state of affairs directly in their physical and social fabric. Landry (2000) has alluded to the same phenomenon in terms of the encompassing notion of the creative city. Some of the most advanced illustrations of this notion can be found in great city-regions of the modern world. Certain areas in these cities display a more or less organic continuity between the local physical environment (as expressed in streetscapes and architecture), associated social and cultural infrastructures (museums, art galleries, theatres, shopping and entertainment facilities, and so on), and the firms that cluster in adjacent industrial districts specializing in activities such as advertising, graphic design, audiovisual services, publishing, or fashion clothing, to mention only a few. Numerous cities have sought to promote this continuity by consciously reorganizing critical sections of their internal spaces like theme parks and movie sets, as exemplified by Times Square in New York, The Grove in Los Angeles, or the Potsdamer Platz in Berlin (Zukin 1991, 1995; Roost 1998). In these cities, work, leisure, and social life increasingly ramify with one another in synergistic interrelationship. The music scenes of Los Angeles and New York dramatically exemplify this trend, with their vibrant mix of live music venues, bars, restaurants, boutiques, and so on, and their associated recording industries (Brown *et al.* 2000; Gibson 2003). The success of these two complexes of specialized urban life can be judged in part by the fact that they consistently turn out streams of hit records in numbers that disproportionately and significantly exceed even those that we might expect from their great size (Scott 1999*b*).

Of course, the cultural economy, along with new-economy sectors at large, is also caught up in insistent processes of globalization, in the twofold sense that (a) producers are more and more inclined to shift relatively standardized work tasks to low-cost locations in other countries, and (b) final outputs flow in ever increasing volumes through international markets (Scott 2002*b*, 2002*a*). Like many other dynamic sectors of modern capitalism, cultural-products industries are dominated by large multinational conglomerates, most of which are involved in all phases of production, from content origination, through distribution, to final sales. The individual majors embedded in these conglomerates typically command significant economies of scale and scope, a circumstance that which permits them to concentrate their creative energies on the production of ambitious blockbuster outputs for global markets (Scott 2002*b*). At the same time (and nowhere more than in audiovisual and media sectors, such as film and television, as already noted in the discussion of Figure 1.6), the majors are engaged in building global networks of creative partnerships such as international joint ventures, strategic alliances, co-productions, and so on. One of the benefits of these arrangements is that they allow producers to scour the world for talent, skills, and ideas. The Hollywood majors are the driving force behind this trend as they push ever forward in the race to produce successful global blockbuster films.

International markets for cultural products are currently dominated by the outputs of majors located in a small number of global city-regions, but they are also increasingly subject to contestation by producers in other places. A number of secondary cultural-products agglomerations around the world, even in peripheral countries, are in the process of upgrading their creative capacities and expanding their presence on export markets. Bollywood cinema is a notable case in point (Pathania-Jain 2001). Many of these agglomerations have useful local assets in terms of distinctive traditions, styles, and world-views, but producers often remain unable to tap into wider markets because the cultural codes in which their outputs are enveloped remain undecipherable to much of the outside world. This state of affairs presents enormous creative challenges to producers and workers in these agglomerations as they seek to maintain significant levels of product differentiation (monopolistic competition again), and yet to cultivate more syncretic sensibilities so that their outputs are able to command a mounting share of global markets.

2.6 COLLECTIVE ORDER OF THE CREATIVE FIELD

The creative field that undergirds all sectors of the new economy, from high-technology manufacturing to cultural-products industries, is constituted as a constellation of workers, firms, institutions, infrastructures, communication channels, and other active ingredients stretched out at varying densities across geographic space. This network of forces is replete with synergistic interactions variously expressed as increasing returns effects, externalities, spillovers, socialization processes, evolving traditions, and so on, and it is above all a locus of extraordinarily complex learning processes and knowledge accumulation. These properties of the creative field mean that much of it functions as a sort of commons, especially in the case of industrial agglomerations where so many of its constituent elements and processes are directly describable in terms of marshallian atmospherics. The same atmospherics often exhibit signs of severe over- or under-production and misallocation, and places where these maladjustments occur are liable to underperform in economic terms relative to their theoretical optimum (Lawson 1999; Niosi and Bas 2001; Oinas and Malecki 2002). There are, then, real gains to be made where systems of public oversight can be brought to bear on these problems of the creative field, though once this statement has been made, it must immediately be qualified by the observation that our current insights and capabilities in terms of relevant policy-making and planning still remain far from equal to the task (Storper and Scott 1995).

In the period extending roughly from the 1920s to the 1970s, the perceived functional failures of industrial innovation systems (and the suggested policies directed to their rectification) were formulated in very different terms from those that inform the present discussion. This was the period when a dominant system of fordist mass production was running its course, and when formal technological research was seen as being *the* critical source of increases in productivity. As economists of the period like Arrow (1962) pointed out, this kind of research was (and is) subject to severe market failure. Private firms have difficulty in exerting ownership rights over any new knowledge that they may produce, with the consequence that they tend to underinvest in research relative to potential overall social returns. Arrow's point, correctly, is that this state of affairs explains and justifies pervasive governmental programmes in support of basic research.

This market-failure problem has by no means disappeared, and government remains—as it must if capitalism is to realize its full growth potential—a major source of research funding and relevant regulatory activity. One important instance of the role of government in this matter is offered by the Bayh-Dole Act of 1980, which has allowed for significant extension of patenting power across different products and institutional arrangements

(Orsi and Coriat 2003). The Act has greatly strengthened the intellectual property rights regime in the United States, with the consequence that a massive surge of new innovative activity has occurred over the last decade or so (Antonelli 2003; David and Foray 2002). In the case of cultural-products industries, intensified attack by international regulating agencies on the problem of product piracy will also greatly fortify the intellectual property rights of producers and presumably help to boost creative energies in that segment of the economy. Further, as the new economy has gathered momentum, numerous shifts in the nature of public support for innovative activity have come about, with non-governmental agencies such as civil associations, foundations, and private–public partnerships acting more and more to complement direct subsidies by the state.

Of course, much public support for innovation also comes from purely local sources. Since the 1970s and 1980s, many municipal authorities and other regional agencies have played an increasingly important part in bolstering agglomeration-specific forms of innovation by providing specialized infrastructures such as research laboratories, technological advisory boards, design centres, and the like (Bianchi 1992; Castells and Hall 1994). Specialized education and training activities subsidized by local governments are also invariably to be found in and around dense industrial agglomerations. In addition, public and quasi-public agencies frequently contribute to the formation of local social capital by promoting trade fairs, exhibitions, festivals, cultural preservation measures, and so on, all of which have considerable direct and indirect impacts on entrepreneurship and innovation. In much the same way, there has been a notable expansion of late in the formation of institutional arrangements for the management of regional trademarks and warranties, certificates of geographic origin, and so on, all of which are important devices for protecting local economic interests. In the new economy, local authorities themselves are increasingly becoming part and parcel of the entrepreneurial and innovative powers of the creative field (cf. Harvey 1989).

The wider urban environment itself also plays a major role in supporting the collective order of the creative field. Urban planners have always been concerned with issues of infrastructure and land use, but today their activities are focused more directly than they ever have been in the past on local business development and the encouragement of innovative industries. A dramatic illustration of these shifting priorities is provided by the efforts of municipalities all over the world to develop science parks in order to encourage local economic development and growth. Some of these efforts—such as Stanford Science Park, Research Triangle Park, or Sophia Antipolis—have been spectacularly successful, though many others have failed to live up to their expectations (Luger and Goldstein 1991). The grandiose Multimedia Corridor Project in Malaysia offers another illustration, with its focus on the promotion of new media industries and related services (Bunnell 2002). A yet further case of an urban planning initiative focused on supporting industrial creativity and innovation can be found in recent transformations of the central garment-manufacturing quarter of Los Angeles. In an effort to upgrade both the local environment and the clothing industry, much of this quarter was turned into a specially zoned enclave in the mid-1990s, officially designated the Fashion District. This enclave now exudes a carnivalesque atmosphere reflecting the renovated buildings, colourful street scenes, and up-scale shopping facilities that have sprung into existence since its creation. Although sweatshop factories still abound in the area, it has increasingly become a centre of innovative fashion design and a unique tourist attraction (Scott 2002*a*). Parallel developments are observable in the planned cultural quarters that can now be found in a number of old European manufacturing cities (Brown *et al.* 2000; Jeffcut and Pratt 2002).

Institution-building to manage the plethora of information flows (and derivative learning effects) in spatial and functional clusters of producers offers further possibilities for collective action in the creative field. We have already noted that these flows often take the form of involuntary and unreliable spillover effects. Institutional arrangements that are capable of at least partially internalizing and rationalizing these flows are therefore highly desirable elements of the creative field (see e.g. Audretsch 2003; Lissoni 2001; Walcott 2002). Experiments in building arrangements of these kinds have proliferated in high-technology industrial agglomerations all over the world in the last couple of decades (Castells and Hall 1994). One outstandingly successful instance is the CONNECT programme in the San Diego area which seeks to promote the local biotechnology industry by linking private entrepreneurs with science and business programmes at the nearby University of California at San Diego (see Scott 1993). In many cases, institutional arrangements like these also function as instruments for engendering trust between the participants in any given local economic system. Trust is a real but elusive lubricant of information flow and hence is an important stimulus for innovation. In the absence of trust—or perhaps better yet a calculated sense of mutual co-dependence—the long-term collaborative interaction that is essential for even-handed exchanges of critical information between private firms can rarely be established (cf. Sabel 1993). 'Calculated' is the operative word here, for an excess of naïve trust only opens producers up to predatory business practices.

Lastly, the status of many segments of the creative field as complex structures of interdependencies suggests that their evolutionary trajectories through time will usually be marked by high levels of path-dependency (Boschma and Lambooy 1999; Nelson and Winter 1982). Path dependency is very characteristic of agglomerated production systems with their interlocking webs of transactions and local labour market activities, which generally give rise to a developmental process that is strongly subject to hysteresis. Institutional mechanisms that provide for at least some sort of collective system-steering can occasionally be beneficial in these circumstances, and, difficult as the task may be, may help to keep the pathway of forward evolution focused on socially desirable outcomes. Two hypothetical cases illustrate the point. First, when a new industry emerges, it will sometimes assume an agglomerated geographic pattern at the outset, and will then rapidly begin to generate localized increasing returns effects. In this manner, the agglomeration as a whole benefits from so-called first-mover advantages, and it will tend systematically to outstrip all later competitor agglomerations by reason of its advanced dynamic of specialized entrepreneurship, innovation, and economic development (cf. David 1985). Given the right conjuncture of circumstances, policy-makers can exploit first-mover logic to help push their region ahead in the developmental race. Second, as any regional economy grows, it may start to lock into increasingly dysfunctional configurations, as in the case of the high-technology firms along Route 128 cited by Saxenian (1994). Lock-in of this sort has many possible sources, but the induration of social and cultural conventions that may once have been beneficial, but then become shackles on further innovative advances, is assuredly one of them. Constant vigilance and mutual cross-checking are therefore required by all stakeholders in order to head off looming problems in this regard.

2.7 GEOGRAPHY AND CREATIVITY

The creative field as identified here is representable as a nexus of multiscalar interdependencies running differentially throughout the domains of production, work, and territory. I have argued at length that attention to this tense force-field of relationships can help us understand a number of critical dimensions of the performance of modern economic systems. I have also suggested that very basic modulations of these relationships occur from place to place as a function of underlying spatial and locational processes. Geography, in other words, is not simply a passive frame of reference, but an active ingredient in economic development and growth. I should add that in my portrayal of entrepreneurship and innovation as socially and spatially embedded processes, there is no attempt on my part to depreciate the role of individual intelligence, imagination, and initiative in the eventuation of these phenomena. I have argued, rather, that human action is always an expression of the integrity of each individual's power to choose and to challenge, even as it is simultaneously and organically situated within real social structures that constitute the basic terrain of action.

As the difficult issues addressed by this analysis have come into view, I have hinted at a further question that can now briefly be made explicit. Why, in short, do certain places at certain times develop as foci of remarkable creativity in the form of exuberant entrepreneurship and innovation? Why did Lancashire become such a prominent centre of the cotton textile industry in the nineteenth century and a vortex of related inventive genius? Why and how did Hollywood emerge as a world centre of film production some time after 1915? Why did Silicon Valley evolve into a hotbed of high-technology entrepreneurial effort and innovation during the 1960s and 1970s?

The analytic deconstruction of the creative field as set forth above points to some fruitful ways of investigating these questions. By way of amplification of this remark, imagine a schematic sequence of events somewhat as follows. Assume that development is initiated in some region by the establishment of a single unit of production, rather like the planting of a seed. For the sake of argument, the precise geographic location of this event may be taken to be entirely random. If subsequent growth of the type illustrated in Figure 2.1 occurs, the evolving industrial system will face many developmental options, but all of them will be marked by path-dependency. Regional competitive advantage will then be secured by an endogenous dynamic of intensifying agglomeration economies in combination with the expansion of external markets. Throughout these stages of development and growth, the active mechanisms of change (i.e. the principal expressions of human agency) are concentrated on the phenomena of entrepreneurship and innovation, which in turn engender modifications in local economic structures, which in turn constitute the field of opportunities over which human action operates, and so on in indefinite sequence. More generally, any attempt to answer the questions posed above must formulate the problem by reference to a dynamic of cumulative causation whose logic is definable not in terms of some *primum mobile* or first cause, but in terms of its own historical momentum. This remark points once more to the importance of an ontology of regional growth and development that is rooted in the idea of path-dependent economic evolution and recursive interaction. By the same token, it also reconnects the discussion to the theories of Marx and Schumpeter, Bourdieu and Giddens, as conjured up at the beginning of this chapter.

One final puzzling question remains. The essence of the question has already been partially articulated above, but I shall now re-frame it in terms that bring it more plainly into view, namely: Is it the quest for enhanced innovative energy that induces firms to agglomerate together in geographic space; or is it the prior convergence of groups of firms around their own centre of gravity that gives rise to the high levels of knowledge creation and innovation so often observed in dense agglomerations? Different analysts veer to different ways of addressing this question, though there is currently an influential school of thought that puts heavy stress on the virtues of learning and innovation as the primary factor in agglomeration (e.g. Brown and Duguid 2000*b*; Malmberg and Maskell 2002; Pinch and Henry 1999; Storper 1995). I want to suggest that there cannot be any cut-and-dried resolution of this issue outside of any given concrete case, and that the question itself is actually not properly posed. Individual industrial agglomerations, in reality, evolve along quite idiosyncratic pathways. There are plenty of agglomerations that are active and growing even though their creative capacities appear to be limited;[6] conversely, we can point to many agglomerations where the main engine of growth seems rather clearly to be their insistent creative vigour. Nor should we forget the (increasing?) number of instances of highly innovative industrial systems that show no proclivity to agglomeration whatsoever. Thus, even if we can offer reasonably plausible abstracted descriptions of the principal forces underlying agglomeration at large, each individual case of clustered economic development will usually reflect a unique combination of general processes and empirical conditions, as already suggested in the classical marshallian approach to the problem. In this interplay of many forces, we ought not to expect any one of the basic marshallian variables to be universally and consistently dominant across all empirical instances. Here, again, we need to stress that the derivative spatio-temporal dynamics are expressed not in a linear pattern of development but in recursive rounds of cumulative causation.

In other words, as I have argued consistently above, the creative field in all its manifestations can never be adequately grasped as a set of 'independent' and 'dependent' variables, but only in terms of structures of direct and indirect interdependence that work themselves out in many different ways in different geographical and historical circumstances.

3

Geography and Development

3.1 THEORIES OLD AND NEW

Theories of regional development and growth have hitherto focused for the most part on situations in the more developed countries of the world. There is no reason in principle, however, why these theories should not also apply—with suitable adjustments—to cases in less developed countries. Certainly, economic theorists of late have increasingly sought to deny that we need radically different approaches for dealing with less as opposed to more advanced economies (cf. Bloom and Sachs 1998; Sachs and Warner 1997). In recent years, indeed, a growing body of empirical work has demonstrated that very similar kinds of regional development and growth processes to those found in North America, Western Europe, and Japan are observable in much of the rest of the world. These processes are manifest in localized industrial systems that range from the purely incipient to large-scale productive regions with global reach.

In the present chapter, I attempt to systematize some of the main theoretical issues that are encountered in any attempt to understand the logic and dynamics of regional production complexes in less developed countries. In addition, I offer a brief review of some of the empirical work that has been undertaken on this question in Asia, Latin America, and Africa, together with some comments on the dilemmas that policy-makers in these areas must face up to in any attempt to promote development.

I proceed at the outset by drawing both explicitly and implicitly on three major strands of thought. The first of these is what Krugman (1996) has called High Development Theory, with its central focus on virtuous circles of cumulative causation and balanced growth. The second is the so-called new growth theory, which emphasizes the pervasiveness of dynamic increasing returns effects in the modern economy (Lucas 1988; Romer 1986). The third is contemporary economic geography, where a long tradition of research has underscored the important role of regional clusters of production and work as motors of economic expansion and social progress (cf. Scott and Storper 2003).

Taken together, these strands of thought provide important clues about effective strategies for the geographic analysis of economic development and especially of the critical stage characterized by Rostow (1960) as take-off, when a society starts to emerge from stagnation into the early phases of industrialization and economic growth. Many less developed countries are caught in vicious circles as represented by low-level equilibrium traps, chronic labour surplus situations, critical shortages of entrepreneurial talent and skilled labour, and so on (cf. Leibenstein 1954; Lewis 1954). In such cases, take-off is unusually hard to achieve, though growth can sometimes be initiated by general push effects that establish a platform for future developmental pathways (Rosenstein-Rodan 1943; Murphy *et al.* 1989). Whatever the initiating factors of take-off may be, a logic of cumulative causation is typically sparked off when industrialization advances beyond some critical threshold level. As this occurs, an intensifying flow of endogenous externalities and increasing returns effects helps to consolidate competitive advantages and to propel development further forward. A degree of spatial concentration of production is correspondingly liable to set in as firms gravitate to particular regions in order to translate latent collective benefits into the realizable form of agglomeration economies. The existence of pervasive spillovers in any region means that the powers of markets to deal efficiently with local resource allocation issues will be deeply compromised. The resulting market failures are especially characteristic of low- and middle-income countries, and market coordination problems are compounded where the institutional foundations of the economic order have not yet begun to mature. In circumstances like these, joint action is an essential accompaniment of regional economic take-off.

These ideas can be summarized more generally in the proposition that development is critically dependent on the formation of dense regional economies marked by strong dynamics of cumulative causation, and can be enhanced by policies providing appropriate forms of collective decision-making and action. The rest of this chapter represents a broad exploration of the conceptual logic and empirical meaning of this proposition, though the reader should keep in mind as we proceed that development is a multidimensional phenomenon, and that the region-based approach advocated here is just one facet of a very much wider set of analytical issues.

3.2 INDUSTRIALIZATION ON THE GROUND: POLEMICAL PRELUDE TO A GEOGRAPHY OF DEVELOPMENT

Let us initiate the discussion by examining an extreme but influential set of normative propositions about developmental processes in low- and middle-income countries. I am referring here to what is frequently designated as the Washington Consensus (cf. Stiglitz 2002). This is a body of economic ideas and political advocacies whose main thrust revolves around claims about the imperative of market organization, the need for clear property rights, and the importance of sound macroeconomic policies (but otherwise limited government interference) in any conscientious effort to pursue development. In this view of things, the market is above all the instrument that will ensure orderly economic growth, because with properly functioning markets, capital and labour will be appropriately mobilized and outputs priced for efficiency.

Before we proceed further, I propose to break the actual process of industrialization down into some of its detailed components, on the ground, so to speak. This will help us, in the first instance, to penetrate beyond the formulaic abstractions of market theory and rhetoric that constitute the core ingredients of the Washington Consensus, and to pinpoint some of the concrete conditions that must be in place before any kind of development whatever can proceed. At a bare minimum, then, the following practical problems need to be resolved in order for a viable system of industrial production to be established and to function. Infrastructural artefacts must be provided and basic public services such as security and education organized; cadres of entrepreneurs have to be on hand in order to set up units of production and to manage their growth; investment funds need be raised, and specific technological and organizational solutions to the problems of production resolved; input–output relations must be put in place and structures of interaction between individual producers and subcontractors set in train; information about wider sales outlets have to be obtained, and customers appropriately cultivated; a labour force with the requisite skills and know-how is required in the vicinity of production units, and some minimal standards of housing and social reproduction need to be promoted in the local community; political collisions between employers and workers must be managed and reasonably viable social frameworks for negotiation of conflicting interests established. In other words, a definite, synchronized sequence of building blocks must be laid out on the ground (i.e. in concert with but in contrast to purely macroeconomic concerns) while simultaneously paying due attention to their functional interdependencies (cf. Amsden 1996). Observe that this thumbnail outline involves an intricate mix of individual initiative and interdependencies together with at least some degree of institutional backup. We might multiply the number of items in the list a hundredfold.

It should be mentioned at the outset that markets, property rights, and macroeconomic order do indeed provide a powerful social context for efficient integration of these system elements and for the effective channelling of capital and labour into useful configurations. We may ask, do they offer all the conditions under which the individual components of an industrial economy will spring forth spontaneously in appropriate substantive form and in proper succession, and come into mutual synergistic relation with one another? We need to keep in mind here that markets, property rights, and macroeconomic order *are not the same thing* as the phenomena that constitute development on the ground. While there must be some sort of balance between the two domains of economic order, the realization of the latter entails logics and behaviours that go well beyond the scope of the former. Of course, some of the basic elements of economic development, such as large-scale infrastructural artefacts, education, and the like, are technically public goods that even many ardent supporters of the Washington Consensus probably recognize as requiring some sort of extra-market supply mechanism. But even when we abstract away from this aspect of the problem, the answer to the question must still be in the negative, for the following reasons.

First, markets, property rights, and macroeconomic order provide critical incentives for industrial development, but they cannot offer necessary and sufficient conditions for the appropriate types of human mobilization to occur, because requisite forms of know-how, practical skill, and acculturation into the norms of capitalist performance must be present too. This problem is compounded in situations where high levels of risk are endemic and where information is both scarce and imperfect. Second, as industrial systems start to develop by means of proliferating interdependencies (between firms, between workers, and so on) they habitually generate meta-market externalities. Atomized competitive economic behaviour is constitutionally prone in these circumstances to lead to market breakdowns and sub-optimal equilibria that can only be corrected by means of collective action. Third, industrialization involves strong path dependencies and circular patterns of cumulative causation—notably in regional contexts—as was well recognized by early theorists like

Gerschenkron (1962), Hirschman (1958), Kaldor (1970), Lewis (1954), Myrdal (1959), Nurske (1959), and Rosenstein-Rodan (1943). Policy intervention is needed to help steer development trajectories away from the low-level equilibrium traps that are endemic in these circumstances. Fourth, and in any case, markets, capitalistic property relations, and macroeconomic order are themselves endogenous to the entire process of industrialization and modernization (cf. North 1998; Polanyi 1944). These phenomena are not independent variables that precede development and then regulate its course, but are one of its contingent outcomes. Above all, and contrary to the faith held in some quarters that competitive pricing mechanisms and smoothly functioning supply and demand systems represent a sort of state of ontological primacy in human society,[1] markets do not come about ready formed. Markets do not emerge *ex nihilo*; neither do they always spring spontaneously into being once incompatible social 'irrationalities' have been cleared away. Even in the presence of massive capital and labour assets, markets only begin to make their historical appearance in response to the formation of appropriate institutional infrastructures and as human expectations are socially rebuilt. The case of Russia after the collapse of the Soviet Union dramatically illustrates this notion. A further telling point may be added by appeal to the infant industry argument of List (1977/1841). One side of this argument refers to the fragility of developing economies in relation to competition from more robust producers, and suggests that concomitant policy measures are required to protect budding entrepreneurial activities. More importantly, the argument points equally to the social externalities of industrial development, and to the practical experience of industrialization as a necessary prerequisite for building skills, know-how, technological competencies, and so on, that would otherwise lie dormant. By the same token, some sort of policy initiative is frequently needed in order to ignite the entire process of entrepreneurship and growth and to maintain the pattern of point and counterpoint that constitutes its inner logic.

It follows that in addition to the role of markets, property rights, and prudent macroeconomic arrangements in stimulating developmental processes, we must insist on the importance of some sort of agency of collective decision-making as a means of activating basic assets, of resolving externality problems, and of social reconstruction, above all in economies at the take-off stage where, precisely, markets are less than fully formed. Small wonder, in view of the manifold difficulties these economies face as they embark on industrialization, that they so readily take to peculiar extra-market coordinating mechanisms-*cum*-conventions like zaibatsu, chaebols, protestant or confucian ethics, extended family and clan relationships, ethnic networks, mafias, and, to be sure, indicative planning and developmentalist political institutions, even if, on occasions, the attempted fusion goes badly wrong. Multinational corporations sometimes play an analogous role to these coordinating mechanisms by mobilizing resources for production in situations where local capacities for entrepreneurship and investment are deficient (Scott 1987). As we might expect, moreover, there is no single formula for success, and many opportunities for failure. Hong Kong is often, and correctly, pointed out as an impressive example of market-led development, but then, Hong Kong's industrialization was preceded by a lengthy historical period of specialized adaptation under British tutelage (and Commonwealth preferences) to international commercial culture and norms of capitalist enterprise.

The World Bank (1993) has claimed that the East Asian experience confirms the importance of 'market-friendly' methods of development. In view of what has just been said, the Bank might more accurately have stated that the experience of East Asia actually reflects a diversity of *market-and-policy-friendly* approaches. Nowhere is this revised formula more pertinent than in the case of those dense agglomerations in the less developed countries of the world, where competitiveness depends crucially on the assets of the urban-industrial environment, but where the associated social, economic, and technical breakdowns are legion.

3.3   CITIES, REGIONS, AND PRODUCTIVITY

Figure 3.1 provides initial, but entirely provisional empirical evidence in support of the idea that spatial agglomeration and economic development are in some way interrelated. The figure shows the relationship between GDP per capita and levels of urbanization for 171 countries, both rich and poor. The strong positive relationship between these two variables is instantly evident in the figure and is confirmed by a very significant correlation coefficient of 0.72. But how, precisely is this relationship structured? What main lines of causality are involved? And what additional variables need to be considered in order to define the

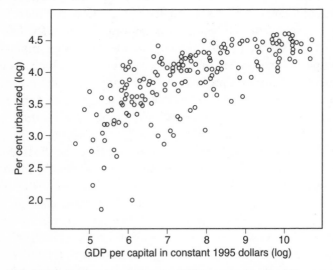

Fig. 3.1. Urbanization in relation to GDP per capita, for 171 countries in the year 2000

Source: World Bank, World Development Indicators.

relationship more thoroughly? I shall start to address these questions by reviewing some of the econometric evidence on the effects of industrial agglomeration on productivity. In subsequent sections, I shall attempt to unpack this evidence into its more detailed conceptual and empirical constituents, though it is worth stating, once again, that relationships of this sort are almost always ones that involve complex recursive processes, in contrast to linear one-way causalities from a set of independent variables to economic growth.

There is a large extant literature stemming from the original statistical work carried out in the 1970s by analysts like Carlino (1979), Kawashima (1975), Shefer (1973), and Sveikauskas (1975) on the productivity-enhancing capacities of cities and regions. Most of this literature uses standard production functions of the form $Q_t = A_t f(K_t, L_t)$ to explore how agglomeration economies combine with capital (K_t) and labour (L_t) inputs to create output (Q_t) at time t. The term A_t is a measure of joint factor productivity, and in the literature that I invoke here, it typically incorporates a proxy measure for agglomeration economies. In this literature, agglomeration economies are then broken down into so-called localization (intra-sectoral) effects and urbanization (inter-sectoral) effects, though as I argued in Chapter 2, these two categories leave much to be desired. They are statistically convenient (because they can be easily operationalized with official census statistics), but they are essentially opaque as theoretical concepts. Even so, the literature provides plausible technical evidence of the widespread existence of agglomeration economies. By far the greater part of the published research on this issue refers to empirical cases in the more developed countries of the world, but there is now, too, a growing body of work that seeks to examine the quantitative effects of agglomeration on productive efficiency in low- and middle-income countries.

Henderson (1986, 1988), for example, has carried out production-function analyses of two-digit industries in the metropolitan areas of Brazil, and has found evidence that industrial productivity is strongly and significantly related to agglomeration. According to Henderson, localization economies play a dominant part in this regard, while urbanization economies are present but only weakly so (see also Henderson and Juncoro 1996; Richardson 1993). Lee and Zang (1998) arrive at comparable conclusions in their study of manufacturing industry in South Korea; they suggest that if employment in any sector in any region were to double, the gross output per worker would increase by 3.0 per cent, and value-added per worker by 7.9 per cent. In a study of Indian cities, Shukla (1988) demonstrates that equally significant increments to productive efficiency are generated by urbanization, and that a massive 51 per cent increase in productivity can be detected as we shift our attention from cities of 10,000 inhabitants to cities of 1 million. The latter findings are backed up by Mills and Becker (1986) and Becker *et al.* (1992), who indicate that productivity advances in Indian manufacturing increase significantly with city size, and by Chen (1996), who shows that agglomeration economies in the food and machinery industries in China are positively correlated with total urban population (though in the case of Shanghai agglomeration diseconomies are apparently detectable). Similar kinds of statistical evidence confirming the impacts of agglomeration on productivity in less developed parts of the world can be found in Fan and Scott (2003), He (2003), Lall *et al.* (2004), Mitra (2000), and Pan and Zhang (2002).

The growing literature on the *intra*-country econometrics of agglomeration and productivity has unfortunately not been matched by an equivalent effort to investigate parallel *inter*-country processes.[2] In view of the paucity of published evidence on cross-country effects, a very modest attempt was made for the purposes of the present study to construct a relevant model based on a standard production function augmented by a term designed to capture the possible impact of urbanization on gross domestic product per capita. Calibration of the model was performed on the basis of data drawn from countries at every scale of development for 1995. The detailed specifications and results of the analysis are laid out in the Appendix, and only the most general conclusions are discussed here. The computed model reveals that GDP per capita for any country is positively and significantly related to (a) the capital/labour ratio, (b) the percentage of the total population living in urban centres, and (c) average years of secondary schooling. As suggested by equation 3 in Table A1 (on page 122) a change in the rate of urbanization in any country from, say, 40 to 60 per cent, will on average be accompanied by a rise of 6.7 per cent in GDP per capita. The impact of urbanization on the dependent variable actually intensifies significantly as we approach the 100 per cent level, a finding that is consistent with rising levels of urban efficiency as we move from low- to high-income countries.

All of the empirical evidence marshalled above points unmistakably to the conclusion that economic growth and rates of urbanization are strongly and positively interrelated in many different parts of the world, including the less economically developed countries. It is probably fair to observe, moreover, that empirically calibrated econometric models are likely consistently to underestimate the influence of urbanization on the formation of national income because there are no actual instances of economic systems where production has advanced beyond agriculture but where cities do not exist. The main problem now is to untangle more precisely the causalities linking urbanization and economic growth together, for if the former has positive impacts on the latter, so—via processes of circular and cumulative causation—growth in its turn is assuredly a stimulus to further urbanization (Renaud 1979). This means that the model presented in the Appendix is less than fully satisfactory because it fails to deal with the inter-temporal relations implied by this observation, though it can be taken as a reasonably useful step in the right direction and as a reference point for further theoretical speculation.

3.4   REGIONAL DEVELOPMENT IN LOW- AND MIDDLE-INCOME COUNTRIES

The Urban Predicament

According to Wheaton and Shishido (1981) rates of urban concentration tend to follow a bell-shaped curve relative to levels of national development. Very poor countries are rather sparsely urbanized, and much of their population lives in rural areas. As per capita income rises, rates of urban concentration go up, and in middle-income countries a distinct tendency to hyperurbanization becomes apparent with the principal city in any given case commanding a disproportionately large share of the total population. With yet further increases in income, the pattern of urbanization becomes less concentrated again, in the sense that the share of the population living in the principal city declines, though the urban population as a fraction of the whole continues to increase asymptotically upwards.

In the light of trends like these, numerous scholars and policy-makers in the late 1970s and early 1980s proclaimed that many Third World countries, and especially those moving into the phase of hyperurbanization, were afflicted by serious diseconomies of agglomeration. Lipton (1977) wrote critically about what he called the 'urban bias' in economic development policy, and, along with a number of other analysts, pointed to the evident overcrowding, congestion, and social breakdowns that seem to be an inevitable adjunct of large-scale urbanization in poor countries. Strong arguments subsequently made their way around various academic and policy communities to the effect that polarization reversal policies would relieve a great deal of the pressure and would in all likelihood actually accelerate processes of national development and growth (cf. Richardson 1980; Todaro 1980; Townroe and Keen 1984). These views were for the most part short-lived, however, not only because they unduly depreciated the productivity gains to be obtained from urbanization, but also because any attempt to implement polarization reversal policies on a significant scale is likely to incur high costs (Storper 1991). Certain kinds of social benefits may well be expected to ensue from vigorous polarization reversal in less developed countries, but in view of the arguments laid out here, reversal would surely undermine the possibilities for accelerated industrial take-off, all the more so given that poor countries can ill afford to spread expensive but indispensable physical infrastructure across wide swaths of terrain (Henderson *et al.* 2001; Mitra 2000).

Note that these remarks are not intended to suggest that developmental efforts on the agrarian, rural front have no useful role to play, or that significant diseconomies never show up as cities or regions grow in size. The point is, rather, that the long-run benefits to urban growth in less developed countries appear almost always to outweigh the costs, and that these costs can usually, in any case, be moderated (given sufficient political will) by suitable kinds of policy intervention and urban planning activities. Since the early 1980s, polarization reversal policies have actually passed quietly from the agendas of the international development agencies that once saw them as an indispensable instrument of economic and social progress.

Agglomeration, Export Orientation, and Growth

In the older version of development theory and practice based on growth-pole and growth-centre analysis expounded by Perroux (1961) and Boudeville (1966), the main avenues to economic advancement were seen above all as passing through capital-intensive industrialization based on mass production in dense regional complexes of economic activity. The propulsive effects flowing from the lead plants at the pinnacle of the mass-production system, in combination with import substitution policies designed to reduce reliance on foreign inputs, were then expected to be the vehicle through which national economic independence

Geography and Development

would eventually be realized. Programmes to promote these outcomes in various parts of the world periphery appeared to achieve reasonable success until about the mid-1970s, at which point mounting setbacks began increasingly to undermine their workability. Complexes of this sort have been described by Lipietz (1986) as centres of 'peripheral fordism', a term that is intended to convey their technological dependency on first-world R&D and the rigid and authoritarian labour relations that prevailed in local plants. In particular, the generally limited domestic purchasing power of the import-substitution countries combined with the intensifying worldwide economic crisis of the mid-1970s made these programmes increasingly problematical as they shifted into the more complex later stages of implementation.

In today's world, export orientation has more or less everywhere supplanted older import-substitution policies as a strategy for promoting economic development. Since the 1980s, policy-makers in low- and middle-income countries have been less and less concerned with the establishment of autarchic national economies than with the search for profitable niches within the global division of labour. This means, as well, that current development practices are generally concerned with a much more diverse group of industries than was the case earlier, when key mass production sectors were seen as being the primary gateway to accelerated growth. Nowadays, it is not uncommon for developing regions to display signs of economic progress even on the basis of small-scale labour-intensive industries, of sorts that were previously thought to be the very antithesis of modernization, like shoes, furniture, or jewellery. In spite of the fact that such industries virtually always merge at their margins into the informal economy, they have proven to be substantial foreign exchange earners in a number of less developed countries, such as Brazil, the Philippines, Thailand, and above all since the early 1990s, China.

With the initiation of industrial development in any region, there is always some positive probability that an associated nexus of agglomeration economies will make its historical and geographical appearance. Depending on the play of interregional competition, these agglomeration economies will tend to boost the potential for additional local growth (though disabling nega- tive externalities will invariably come into being at some stage if policy-makers fail to act). As we saw in Chapter 1, the main factors underlying agglomeration revolve in complex ways around the spatial structure of transactions costs and external economies of scale and scope. External economies, in turn, break down into network benefits, increasing returns in local labour markets, and the learning effects that tend to flow through dense communities of firms and workers. These inducements to agglomeration become even more potent when they begin to function jointly in a dynamic logic of circular and cumulative causation. To take just one example, the many-sided innovation processes that so commonly operate in dense transactions-inten- sive industrial complexes help to accelerate rates of growth, and continued growth creates further conditions (such as extensions of the division of labour and intensified learning) under which the probability of new innovative occurrences is sharply increased (cf. Lucas 1993; Stiglitz 1989).

Positive outcomes like these are significantly enhanced when export-orientation strategies are in place. No matter what sectors or types of firms comprise the local economic base, export orien- tation enables producers to tap into a vastly extended pool of purchasing power. This in turn makes it possible for regional production complexes to expand to levels that exceed the size that could be supported by a purely domestic market, thereby promoting intensification of localized increasing returns effects and competitive advantages. That being said, one of the big problems developing areas face in this regard is to find and main- tain outlets in the global economy that are not already dominated by producers with superior competitive advantages (due to scale effects or an early start, for example), and this problem is aggra- vated by the crowding out that is currently taking place on global markets for many types of low-cost manufactured products (Palley 2003). The great expansion of the sales of Chinese-made products to the rest of the world since the late 1980s has had particularly harmful impacts on the export opportunities of a number of other less developed areas (Scott 2005*b*).

The export capacities of industrial regions in low- and middle-income countries are often significantly enlarged as a consequence of the activities of foreign-owned branch plants. What is more, a swelling body of empirical evidence shows that many kinds of foreign direct investments generate positive spillovers in receiving regions (cf. Lall 1980; McAleese and McDonald 1978; Rasiah 1994; Roberts 1992; Scott 1987). A now somewhat faded 'corporate imperialism' school of thought,[3] argued strenuously that few benefits could be expected to flow from the activities of foreign-owned firms in peripheral countries, and that the relationship between the two is doomed to be one of parasite and host. Whatever objections on other grounds one may have to the ways in which multinational corporations operate in these countries, the claim that they never bring benefits to their local economic environment can scarcely be one of them. As foreign branch plants spread their roots, they often help to generate new entrepreneurial ventures in the local milieu, combined with intensified flows of technological information, improvements in labour skills, and so on, as well as increases in employment. In a word, foreign-owned branch plants can be a significant source of agglomerated growth and an important medium through which selected regions in less-developed countries find it possible to participate in global commodity chains (Bellak and Cantwell 1998; Christerson and Lever-Tracy 1997; Gereffi 1995; Henderson 1989). Even a number of large African cities are now becoming more deeply integrated into the global economy as foreign direct investment proceeds. Accra, for example, is a major centre of 655 foreign companies acting as intermediaries between the local and the global economies (Grant 2001). In addition, many large buyers (e.g. retailers such as Marks and Spencer or Tesco) in more economically advanced countries have discovered the advantages of setting up global sourcing networks with offshoots that reach deeply into far-flung agglomerations of producers in less developed parts of the world (Crewe 2004; Humphrey and Schmitz 2002).

Yet even when powerful relationships linking industrial agglomeration and the extension of markets together are strongly in evidence, the peculiar market failures and other breakdowns that beset industrialization in less economically advanced parts of the world always threaten to disrupt the virtuous circle of growth. A special set of problems in take-off situations derives from the difficulties of ensuring a reasonably balanced intra-regional pattern of development. It is above all important in emerging agglomerations to ensure that adequate forms of industrial specialization and complementarity are in place. We may perhaps borrow here from the ideas of Rosenstein-Rodan (1943) about the merits of a 'big push' in the guise of central government programmes to ensure balanced forward development. A geographical perspective suggests the possibility that local authorities, for their part, can stimulate 'regional push' effects by encouraging the formation of positive externalities and increasing returns that would otherwise tend to be absent or undersupplied (Scott 2002c). This notion of regional push, it may be noted, runs contrary to advocacies that preach the need for an equal spatial distribution of productive assets in the national economy as a whole. As Hirschman (1958) expresses the matter, the emergence of a small number of 'one or several regional centres of economic strength' is an essential prerequisite for any kind of advanced development to occur. We shall return to the policy issues raised by this proposition later.

3.5 REGIONAL PUSH IN PRACTICE

Low- and middle-income countries present a medley of faces, and generalization about them is fraught with hazards, except perhaps to say that excessively large segments of their populations live in dire poverty. Their economic structure varies greatly from country to country as well as from region to region within individual countries. Various parts of the world periphery have little or no industrial development whatever, and have little prospect of achieving significant economic take-off at any foreseeable time in the future. Other parts are at various stages along the road to development, ranging from early manifestations of industrialization in the guise of small-scale craft activities focused largely on local markets to advanced regional complexes that have achieved virtually full-blown participation in global networks of trade.

Small-Scale Artisanal Industries in Asia, Latin America, and Africa

Small-scale artisanal industries are of special theoretical and practical interest in less developed countries in view of their low entry barriers and labour-intensive modes of operation. Often enough, these industries form simple embryonic clusters based on indigenous traditions and skills. But given the right conjuncture of circumstances, the same clusters sometimes evolve into large-scale agglomerations with important international ramifications.

Asia is a hive of traditional industrial districts, many of which have attained a significant presence on global markets (cf. Cadène 1998; Cawthorne 1993, 1995; Chari 2000; Kattuman 1998; Nadvi 1999*a*, 1999*b*; Sandee 1994; Scott 1994; Tewari 1999). In India—especially after the turn to economic liberalization in the late 1980s—artisanal industrial districts have developed and grown at a rapid pace. Cawthorne (1993, 1995) describes the emergence of a vertically disintegrated cotton knitwear industry in Tiruppur, which has now become a major exporter of T-shirts, though it is still evidently enmeshed in a low-road trajectory of development based on depressed wage levels and limited skills. Chari (2000) refers to Tiruppur as a case of 'capitalism from below', that is, as a case of industrialization based largely on peasant-worker entrepreneurs. Knorringa (1996, 1999) provides a dense account of the traditional shoemaking cluster of Agra where some 60,000 workers are employed in 5,000 manufacturing units. For the most part, the Agra cluster is a low-quality, labour-intensive industry, much given to cut-throat competition and conflict-ridden industrial relations. It is also characterized by a large subcontract sector in which sweatshops abound, along with piecework and homework labour-contracting arrangements (see Waardenburg 1993). Social networks are an important adjunct to the functioning of the Agra cluster, but these display economically dysfunctional cleavages around caste and class. Despite these failings, some 25 of the top producers in the cluster, as Knorringa shows, have achieved a level of quality sufficient to enable them to export to markets in Europe and the United States. Symptomatically, these higher-quality producers evince a relatively strong propensity to inter-firm collaboration. Thailand is another example of a country in which clusters of artisanal industry abound. The traditional gem and jewellery industry of Bangkok, for example, has evolved from a small collection of workshops serving local markets in the 1970s to the major generator of exports that it has become today (Scott 1994). The industry has aggressively carved out this position for itself on the basis of its low wages, its vibrant agglomeration economies, and the skilful political manoeuvring of its representatives. Of all such cases of regional efflorescence in Asia based on labour-intensive, artisanal industrialization, the case of southern China is without doubt the most dramatic (Christerson and Lever-Tracy 1997; Fan and Scott 2003). Here, foreign capital and indigenous entrepreneurship have combined together to create an industrial juggernaut producing huge quantities of textiles, clothing, shoes, furniture, and so on in specialized industrial districts for world markets.

Latin America, too, is well-endowed with traditional industrial clusters making products such as clothing, knitwear, shoes, ceramic tiles, metallurgical products of all varieties, and so on (cf. Altenburg and Meyer-Stamer 1999; Lawson 1995; Meyer-Stamer 1998; Rabellotti 1999; Rabellotti and Schmitz 1999; Villarán 1993; Visser 1999). The research of Schmitz (1995; 1999a; 1999b; 2001) on the shoe-manufacturing agglomeration of the Sinos Valley in southern Brazil represents a particularly accomplished and detailed body of work on this topic. Schmitz shows how this agglomeration of vertically disintegrated firms, whose origins stem from the local availability of hides and leather, has gradually evolved over the last couple of decades from a low-grade supply system selling strictly on the domestic market to a centre of medium-quality women's shoes with burgeoning export sales. Schmitz argues that geographic clustering, a dense texture of positive externalities, and the formation of institutions promoting beneficial forms of joint action have been critical to the success of the cluster, though signs of exacerbated competitiveness have become evident of late. A further point advanced by Schmitz is that as the industrialization of the Sinos Valley has moved forward, so traditional or ascribed forms of interpersonal trust based on kinship or ethnic relations have gradually given way to what he calls 'earned trust' as a basis for business relations. Another way of expressing the same idea, perhaps, is to say that as traditional industrial clusters become more thoroughly imbued with the spirit of modern capitalism, the business conduct of individual owners and managers tends on balance to become more reliable, thus opening up new possibilities of collaboration and interdependence. The maintenance of collaborative working relations is especially important in view of the strong disposition to opportunistic and free-rider behaviour that has been noted in industrial clusters where the pressures of day-to-day existence undermine more long-term horizons of calculation (van Dijk and Rabellotti 1997; Knorringa 1996). Parallel findings to those of Schmitz have been reported for the shoe industry in Mexico by Morris and Lowder (1992) and Rabellotti (1997, 1999).

Africa, which contains an unduly large proportion of the world's poorest countries, has a number of regional clusters of artisanal industries, though they are generally less in evidence and less advanced than they are in Asia or Latin America. This state of affairs is probably not so much because Africa is an erratic case in terms of its inherent developmental logic but because, for much of the continent, the circumstances of history and geography have set the starting position so much lower than elsewhere. Studies by

Dawson (1992), McCormick (1999), Sverrison (1997), and van Dijk (1997), among others, have examined a number of artisanal clusters in Ghana, Kenya, and Zambia, with diverse specializations in industries such as fish processing, garments, metalworking, and furniture. These clusters produce mostly low-quality outputs based on unskilled, low-wage labour, and they remain in a largely nascent phase of development. Even in these cases, signs of increasing organizational complexity are observable, ranging from evolving divisions of labour to a distinct capacity for innovation as manifest in the adaptive behaviour of producers faced with the need to make do with recycled inputs and second-hand equipment. One of the more striking of these clusters is the vibrant vehicle-repair industry in Kumasi, Ghana, where large numbers of small-scale workshops provide a profusion of customized and semi-customized services based on the re-fitting of old parts (Dawson 1992). The Kumasi cluster has developed to the point where it now engages in a thriving export trade with surrounding countries in West Africa.

Technology-Intensive Industries

In addition to, and often alongside these mostly traditional types of small-scale artisanal industries, many kinds of technology-intensive industrial agglomerations are in evidence in selected parts of the developing world. In some cases, we can trace their origins back to long-standing craft communities; in other cases, they have either grown spontaneously or have developed in response to foreign direct investment and other foreign entrepreneurial activities; in yet other cases, they are the result of actions carried out by the developmentalist state.

A noteworthy example of a radically upgraded craft cluster is recorded by Nadvi (1999*a*, 1999*b*) in his study of the surgical instruments industry in Sialkot, Pakistan. Nadvi shows how traditional metalworking in Sialkot, formerly a centre for the production of knives, swords, spears, razors, and so on, has evolved into a dynamic agglomeration of firms producing stainless steel surgical instruments. These firms have succeeded in capturing 20 per cent of total world exports in surgical instruments. Family ties and traditional trade associations remain active principles of social organization among producers in the Sialkot cluster, but they appear to have evolved in ways that make them useful rather than obstructive adjuncts to modern business. Nadvi goes on to indicate that, in 1994, in response to restrictions by the US Federal Drug Administration on imports of surgical instruments from Sialkot, firms within the cluster made a concerted effort to upgrade their activities, improving quality standards throughout the supply chain, and rapidly recovering lost markets. Upgrading programmes in general have now become important items on the agendas of local economic development agencies in low-and middle-income countries (Schmitz 2004).

Among the most advanced cases of new industrial spaces in Asia and Latin America are a number of large metropolitan areas with burgeoning complexes of electronics, computer, and software activities. These complexes are not only important foci of economic expansion as such, but also of wider processes of modernization and social change generally (cf. Armstrong and McGee 1985). Their growth at favoured locations is based on (a) a modern sector or sectors, (b) a locally supportive political environment, (c) an abundance of urban amenities, (d) large supplies of suitable labour, and (e) the successful incorporation of local producers into the international division of labour and diffusion of their outputs on international markets. Obvious pioneers of this model are the Asian city-regions of Seoul, Hong Kong, Taiwan, and Singapore, which have grown by leaps and bounds over the last couple of decades on the basis of their technology-intensive industries (Clark and Kim 1995). A case that has received much attention recently in both the press and academic literature is the software production complex of Bangalore in southern India. Since its origins in the 1980s as a centre of so-called body-shop operations offering cheap, short-term subcontracting services (based on the local availability of low-wage technical labour) to US corporations, Bangalore has evolved at a rapid pace, and has steadily moved up the price and quality curve to the point where it has today become a major node within the entire global software industry (Audirac 2003; Saxenian 2002). A marked diversification of high-level development is also occurring in many cities, as illustrated by Kuala Lumpur in Malaysia, which has managed to become (after Hong Kong) one of the main commercial and financial centres of East and South-East Asia (Morshidi and Suriati 1999), and is now, by means of the Multimedia Super-Corridor Project, bent on turning into a pan-Asian focus of electronic media production.

The Balance Sheet

The experience of regional development in low- and middle-income countries represents something of a mixed bag, even though the selection of examples presented above is strongly biased towards more successful cases. On the one hand, much of the industrial activity that takes place in the less developed world consists of traditional craft enterprises operating at relatively low levels of productivity and competitiveness. Even if we can point to notable instances of regions within which some upgrading of these enterprises has occurred, there are many more cases where more advanced development remains only a remote prospect. In some cases, development has not only been truncated but actually reversed, a circumstance that is dramatically illustrated by the Filipino shoe industry, which has gone into deep crisis since the mid-1990s as a result of Chinese competition (Scott 2005*b*). On the other hand, numerous regions in the global South have evolved to the point where it is doubtful if the appellation 'less developed' can still be appropriately applied to them. On the basis of aggressive export-oriented industrialization programmes, many of them have effectively joined the ranks of the more economically advanced parts of the world and are able to compete internationally as much on the basis of product quality as on price.

It must also be recalled that if the formation of dense regional production systems and associated urban excrescences in low- and middle-income countries has a positive impact on development, it is also associated with significant countervailing disadvantages. Among the heavy costs of this form of development are problems of poverty, inequality, crime, congestion, environmental degradation, and so on (Douglass 2001; Stren 2001). Moreover, labour relations in industrial districts in virtually all parts of the less developed world are frequently underregulated and conflict-ridden, if not on occasions positively despotic. Notwithstanding cases where definite improvement of remuneration levels and labour skills has occurred, large segments of almost all these districts are marked by low wages, poor working conditions, casualization, child labour, and wholesale feminization of unskilled jobs (Baud and de Bruijne 1993; Lawson 1995). Additionally, we must acknowledge the fact that foreign branch plants are not always an unmixed blessing in less developed areas, and that it is far from unusual to find cases where they have engaged in serious violations of wider environmental, labour, and fiscal norms. Urbanization and industrialization in less developed countries invariably come at a high price, and strenuous policy intervention is essential to mitigate their negative side-effects.

3.6   INSTITUTIONS AND MARKETS: DEVELOPMENTAL PRACTICE IN REGIONAL CONTEXT

The Regional Economic Commons

As I have already intimated in Chapter 2, spatial agglomerations of productive activity are distinguished by what we might call a *regional economic commons* constituting the kernel of localized competitive advantages. This phenomenon is composed of all the externalities and increasing returns effects actually or latently available to producers in any given region as a result of the co-presence of interrelated firms together with a local labour market and an overarching system of norms, conventions, cultures, and so on. It is, by the same token, the main locus of regional push effects, and it is of particular significance in economies at the take-off stage of development.

The commons benefits all, but is the property of none. It is by definition only partially susceptible to coordination by market forces, and in the absence of management by other means, its benefits are apt to be misallocated or underprovided or both. The situation is made more complicated by the path-dependencies that typically structure the evolution of any regional economy. For this reason alone, what we are liable to find in any given regional cluster at any given moment in time is less a sequence of instantaneous general equilibria than conjuncturally local outcomes whose structure is in important ways a function of previous system histories. As a result of these market imperfections, positive social payoffs are apt to materialize whenever mechanisms of strategic political choice are put into place, hence making it possible to steer the local economy away from less and into more desirable long-run outcomes. Confirmed anti-dirigistes such as Krueger (2000) or Lal (1983) would no doubt object at this point that it is always better in practice to live with market failures than with the 'inevitable' gaffes of public intervention. No matter how salutary this warning might be, it can only be sustained if the phenomenon that I have identified as the regional economic commons is purely illusory, and/or if local decision-makers are irremediably incompetent or corrupt. In any case, for low- and middle-income countries, and even in the absence of technical market failures, the magnitude of the problems they face allows them neither the luxury nor the time to wait for business as usual to take its course. The economic history of countries like Germany, Japan, or Singapore demonstrates that the bases of competitiveness can indeed be socially and politically reconstructed to serve desired developmental goals, and that bold public action is of particular value in the early phases of industrialization.

The virtues of markets, of course, need to be taken very seriously in take-off situations, though it bears repeating that the market itself is part and parcel of the overall developmental process. Hence, and especially at the regional level, a finely balanced and mutually sustaining mix of emerging market relations and institutional order is indispensable for growth, though clearly not just any institutional order will do. Depending on their precise design, institutions can significantly promote or significantly hinder development, and hence issues of institutional quality (relevance, transparency, accountability, flexibility, competence, etc.) call for careful attention (Rodrik 1999). Among other things, institutions need to be sensitive to local idiosyncrasies, and they need to be continually readjusted as the economic system (both local and national) evolves through time.

The regional economy, then, is definable as a collective entity in the precise sense that it is a domain of externalities and competitive advantages in which the destiny of each individual producer is intimately linked to the destiny of all. Concomitantly, the regional economy is necessarily a social and political construction as much as it is an expression of atomized competitive relations. In these circumstances, the point of development policy is less to concentrate unidimensionally on the creation of well-lubricated markets than it is to forge concrete competitive advantages based in the shared order of the economic commons (cf. Amsden 1997). Hirschmann (1958: 5) offers support for this remark when he writes: 'Development depends not so much on finding the optimal combinations for given resources and factors of production as on calling forth and enlisting for development purposes resources and abilities that are hidden, scattered or badly utilized'. And in a similar vein, Schumpeter (1934: 64) points to the creative destruction inherent in all economic growth, and its equilibrium-unsettling dynamic: '...development in our sense is a distinct phenomenon, entirely foreign to what may be observed in...the tendency to equilibrium. It is spontaneous and discontinuous change...disturbance of equilibrium...which forever alters and displaces the equilibrium state previously existing'.

None of these statements implies that it is advisable simply to neglect the efficiency-seeking properties of markets in efforts to promote development, but each in its different way suggests that developmental outcomes are likely to be consistently improved where complex extra-market initiatives are in place. In regional contexts, in particular, proactive policy pushes are essential to stimulate take-off and to secure the foundations of joint economic welfare. The deepening trend to globalization makes the latter point all the more emphatic, for open markets and free trade bring considerable threat as well as opportunity to rising regional clusters.

Policy Instruments for Regional Development in Low- and Middle-Income Countries

The concept of the regional economic commons (which includes, but goes beyond, the notion of the creative field as laid out in Chapter 2) points at once to a number of potentially powerful policy instruments that can be deployed in the search for increased competitiveness. In less developed countries, low-cost policies that actualize latent agglomeration economies by means of bottom-up strategies are of special significance in this regard. At the same time, the goals of policy need to be realistic in the specific sense that public authorities are well advised to proceed prudently on the basis of what already exists in the way of regional assets, and cannot be expected to conjure miracles out of thin air. Three very specific lines of policy intervention can be immediately identified, each of them responding in one way or another to aspects of the collective order of regional production systems.

Industrial networks and collaboration   In less-developed partsof the world, as in many more developed countries, local inter-firm transactional relations tend to be aggressively competitive, unreliable, and unduly devoid of mutually useful information content. These problems are liable to be especially evident in the early phases of take-off when firms are much given to opportunistic behaviour. It is assuredly difficult for policy-makers to alter circumstances like these, but certain forms of remedial action have been shown in practice to provide some reprieve. For example, a number of regional authorities in different countries have found it possible to improve levels of inter-firm trust and reciprocity by setting up forums where leading firms and their subcontractors can discuss problems of mutual concern and work out more effective modalities of interaction. Among other related activities, specially trained brokers can help to promote inter-firm connections and to moderate free-rider problems within regional production networks.

Labour issues   A lack of appropriate skills is almost always a problem in industrial districts in countries at every level of development, but the problem is especially acute in low- and middle-income countries where basic education and training facilities are usually quite deficient. In these cases, public investment in the upgrading of the labour force is a major requirement for further development and growth. Programmes of this sort, moreover, need to be resolutely alert to practical matters of direct relevance to agglomeration-specific needs. Additional problems revolve around the multiple breakdowns that occur in the circulation of useful labour-market information and around the costs that this state of affairs imposes on employers and job seekers alike. Accordingly, some form of institutional support for the tasks of gathering and diffusing information on labour-market conditions is highly desirable. Yet more complex dilemmas arise with respect to the socialization and habituation of the labour force, so that the wider system of social services is strongly implicated in local economic development issues.

Learning and innovation   The status of any region as a nexus of learning and innovation effects can almost always be improved by selective public expenditures on appropriate research and technology-enhancing activities. In regional contexts, above all in less developed areas, basic research is generally less urgently needed than relatively simple services offering technological advice to individual firms, and assistance with the solution of practical problems. Many different kinds of institutional frameworks can secure goals of this sort, including local schools, colleges, and specialized governmental agencies. In some circumstances it will be advantageous to invest public money in an innovation centre where technical issues of local interest can be subjected to expert scrutiny. As Bellak and Cantwell (1998) have pointed out, too, appropriate policy measures can induce foreign-owned branch plants to augment the flow of beneficial technological spillovers to other firms in the surrounding area. In any case, the central policy goal here is to achieve some degree of local economic upgrading and to build competitive advantages above and beyond the mere cheapness of labour.

All of these suggested lines of policy action call for careful institutionalization, and all require thoughtful design in ways that are sensitive to local context. But the possibilities of creative institution-building do not cease at this point. Auxiliary types of intervention are needed to combat many of the other market failures and allocative inefficiencies that are virtually ubiquitous in regional economies in less developed countries. Marketing and export organizations, publicly supported exhibitions and trade fairs promoting local products, and industry associations that impose fiduciary standards and fair practices, represent a few examples of the specialized services that can be usefully offered by the collectivity in these situations. In addition, public institutions supplying credit to micro-enterprises on favourable terms are likely to be particularly beneficial in less developed areas, and have been identified as being of special importance in promoting entrepreneurship among women (Dignard and Havet 1995). Obviously, as well, physical infrastructure and associated urban planning initiatives are critical to the emergence of an orderly local economy. Judicious planning is essential to deal with the diverse diseconomies that always make their appearance in large cities, and that are especially severe in developing areas. Equally, well-equipped industrial parks, export-processing zones, and special economic districts with good access to a local labour supply can be magnets for significant inward investment, foreign as well as domestic. Facilities like these have provided foundations for the growth of major industrial clusters in a number of less developed areas in all parts of the world. The island of Mauritius has been hailed as one of the most outstanding recent examples of a country that has prospered greatly on the basis of an imaginative development programme centred on the establishment of export-processing zones (Roberts 1992).

The benefits that may be expected to accrue to any region as a result of these different kinds of public action will typically be enhanced if some overall mechanism of intra-regional harmonization is established, particularly where different programmes have a tendency to work at cross-purposes with one another. A regional agency or development coalition, no matter how rudimentary its structure, will frequently prove to be indispensable as a locus of system oversight. Agencies with wide responsibility for coordination can on occasions help to steer regions through critical decision nodes aligned along their path-dependent evolutionary course. The infant-industry problem exemplifies the point well. Public support for emerging industries that promise significant gains in the future, but that would atrophy if left to themselves in the present, can be expected to pay dividends in the long run as these industries grow and mature (Lall 1990; Lee 1997). Regional associations, both governmental and civil, can play a significant role in animating public discussions about questions of development and growth, and beyond that, in the promotion of new forms of regional consciousness and identity (cf. Gerschenkron 1962). The case of the ABC region in Brazil dramatically exemplifies some of the main issues here (Scott 2001*a*). Local policy-makers in the region have recently sought to establish an ambitious programme of economic transformation, and to promote the social and political conditions without which further development is likely to be obstructed in various ways. In pursuit of this goal, they have attempted to increase democratization of municipal decision-making and government, while simultaneously seeking to mobilize the local population by means of open debates on the future of the region and on appropriate courses of policy intervention (see also Campbell 2001; Klink 2001).

Success in the matter of local economic development in low- and middle-income countries, as we have seen, is liable to create dense islands of relative prosperity in a wider ocean of impoverishment. The formation of growing industrial agglomerations in less developed countries is invariably accompanied not only by patterns of unequal physical development, but also by persistent interregional income inequalities (Williamson 1965). Exacerbated income inequalities in turn are always liable to lead to political tensions that can directly and indirectly damage any developmental dynamic based on the privileged growth of only a handful of regions, especially if the dynamic is sustained out of the public purse (Scott and Storper 2003). Hence, the ultimate workability of any given region-centric development model may well turn out to be politically non-viable over the long run in the absence of compensating policies. Some spatial reorganization of basic productive assets may be necessary here, though at the risk of much strain on the wider goal of economic development. Perhaps the best approach in these circumstances would be to put some sort of direct income redistribution policy into effect. This would no doubt reduce investment rates—and hence growth—in core regions, but might be expected to induce some offsetting benefits by raising overall domestic demand.

3.7 A WORLD OF REGIONS

I have argued that much important new light can be thrown on development theory and practice by taking the regional question seriously. This proposition holds for economies at every level of per capita income, but it is especially pertinent to the case of economies poised at the stage of take-off where resources are scarce and competitive advantages are usually at a low ebb. I have suggested that a market-and-policy-friendly approach offers the best line of attack on economic backwardness, though finding exactly the right mix of arrangements to fit any concrete situation obviously presents enormous challenges. Blunt boilerplate approaches are certainly unlikely to be successful in any long-run perspective (Storper and Scott 1995). The approach to development that I have broadly sketched out here allows for—indeed necessitates—a wide diversity of remedial actions tailored to the detailed peculiarities of local agglomeration processes. It involves, into the bargain, a process of region-based creative self-discovery and social transformation so that with the passage of time—recalling Hirschman as cited earlier—there is some likelihood that unsuspected local resources, talents, and potentialities will continually be discovered and mobilized.

As export-oriented industrialization programmes have come to play an ever-larger role in national development across the world, their success has at least in part been dependent on the existence of vibrant regional production systems offering unique pools of concentrated competitive advantages. In the same manner, export orientation has allowed a great many less-developed areas over the last couple of decades to launch their products on global market niches, and then to use this point of entry as a means of moving on towards higher-quality, higher-skill production. This is the route followed by the knitwear industry of Tiruppur, the surgical instruments industry of Sialkot, the Bangkok gem and jewellery industry, the shoe industry of the Sinos Valley, the electronics industries of Singapore and Taiwan, and hosts of other cases in low- and middle-income countries. Partly as a corollary of these remarks, the argument presented here points ultimately in the direction of a global economy that is in significant ways constituted as an ensemble of local economies scattered over both less and more developed countries.

By the same token, the old postwar international order with its developmental geography rooted in a core–periphery system seems more and more to be giving way to a new geography in the shape of a global mosaic of regional economies imbricated within a slowly shrinking expanse of underdeveloped territory. If this analysis is correct, it suggests that numerous urban areas on the current margins of world capitalism will eventually accede as vigorous nodes to the expanding global mosaic. Places like Seoul, Taipei, Hong Kong, Singapore, Mexico City, São Paulo, and others, have already moved far along this developmental pathway. Many others now appear to be in the early phases of a similar apotheosis.

In spite of these optimistic comments, there are many areas on the spatial margins of contemporary capitalism where development and growth remain stubbornly elusive goals, and where even the most elementary forms of industrialization are at best a distant prospect. In the new global order that is now emerging, there can be no question, on either practical or ethical grounds, of simply abandoning these left-behinds to their fate. If any meaningful notion of a global community of regions and nations is eventually to be achieved then greatly intensified programmes of aid to such areas will need to be put into place. Equally, the resurgence of new regional economic and political forces in both the more advanced and less developed countries of the world, together with the difficult dilemmas that they engender, suggest that an overarching decision-making framework ensuring some sort of interregional coordination across the globe will become steadily more essential. I have argued elsewhere (Scott 1998*b*) that one of the vital tasks that any equitable version of globalization will need to face in the future is exactly this issue of the governance of interregional relations, irrespective of the geography of national boundaries. The point must be stressed with particular vigour given that the dynamics of development and growth, as laid out here, seem everywhere to be calling forth new forms of region-based political identity, activism, and competition.

Appendix

A cross-country production function was computed in the attempt to decipher something of the relations between urbanization and overall economic productivity. The tentative nature of this exercise must by emphasized at the outset. I should also state immediately that I am well aware of the Cambridge critique of standard production-function analysis and of the concept of aggregate capital (cf. Fine 2003; Harcourt 1972). My objective here is not to make arguments about matters of value and distribution, but to search for regression equations (essentially, engineering functions) that offer some crude inductive sense of the quantitative regularities linking urbanization and gross domestic product per capita.

The dependent variable in this analysis, Q_i, is GDP per capita in constant 1995 dollars for country i. The independent variables are defined as follows:

K_i Gross fixed capital for country i.
L_i Total labour force in country i.
U_i Proportion of total population in country i living in urban centres.
S_i Average years of secondary schooling for the total population in country i.

Data for all variables except S_i were taken from the World Development Indicators database available from the World Bank. All of these data refer to the year 1995 (so as to maximize the number of observations). Data for S_i were taken from the Barro and Lee database on international measures of educational quality which is accessible on the World-Wide Web at http://www.worldbank.org/research/growth/ddbarle2.htm. The variable S_i is defined for the year 1985 (the latest year for which data are available). K_i was calculated from annual data on fixed capital formation going back in all cases at least twenty years, and, where data are available, as far back as 1960. Depreciation was defined at the rate of 15 per cent per annum.

A basic production function was then established with the general form $Q_i = A_i(K_i/L_i)^\alpha$ where $A_i = \text{constant} \times \exp(\beta U_i^2 + \gamma S_i)$. Note that the urbanization effect is expressed in this equation as the square of U_i, a manoeuvre that is intended to compensate for the fact that U_i is defined over a closed number system, with a consequent compression of actual values as U_i approaches unity.

The main computational results of the exercise are laid out in Table A1. The capital–labour ratio performs well in the analysis (equations 1 to 3), as we would expect, though the estimate of α is presumably higher than it would be if the dependent variable were defined in terms of value added per worker rather than GDP per capita. The urbanization effect is positive and significant, as is the effect of the education variable. Equation 4 shows the relation between GDP per capita and urbanization in the absence of the other independent variables. An attempt was made to modify equations 1, 2, and 3 by adding a further independent variable, represented by total population. This manoeuvre was carried out in search of Verdoorn effects, but with negative and non-significant results.

Table A1. Regression parameters

Parameter	Equation			
	1	2	3	4
Constant	2.8261	3.4941	4.1723	645.4833
α	0.9641**	0.9190**	0.8695**	—
β	—	0.3964*	0.3244*	0.4860**
γ	—	—	0.1384**	—
Adjusted R^2	0.96	0.96	0.97	0.53
Number of observations	73	73	66	177

Notes:
* significant at the 0.05 level.
** significant at the 0.01 level.

A finding of some interest in the present context is that any increase in the rate of urbanization from, say, U_1 to U_2 will be associated with a proportional increase of GDP per capita of $e^\lambda - 1$, where $\lambda = \beta(U_2^2 - U_1^2)$. Hence, with reference to equation 3, a change in the rate of urbanization from, say, 40 to 60 per cent in any country will be accompanied by a change of 6.7 per cent in GDP per capita.

Unfortunately, cross-country equations of the sort presented here are typically plagued by the dual problem of collinearity and ambiguity as to the directions of causality. The latter problem is especially acute. Do high levels of the capital–labour ratio, urbanization, and schooling actually cause high levels of GDP per capita, or vice versa? Any sensible answer to this question will presumably insist on two-way recursive directions of causality, as suggested by the theory of cumulative causation. At best, then, the equations laid out here only go halfway towards a meaningful analysis. At worst, they at least indicate that the hypothesized relationship between urbanization and growth is not, for the moment, disconfirmed.

I should add that an attempt was made to compute a cross-country growth accounting model incorporating an urbanization variable. However, this attempt failed to produce any significant results, an outcome that can perhaps be attributed to the combined effects of the empirical heterogeneity of the data and the circumstance that rates of urbanization change slowly relative to changes in national income.

REFERENCES

Acs, Z. J. 2002. *Innovation and the Growth of Cities*. Cheltenham: Edward Elgar.

—— L. Anselin, and A. Varga. 2002. 'Patents and innovation counts as measures of regional production of new knowledge'. *Research Policy* 31: 1069–85.

Allen, G. C. 1929. *The Industrial Development of Birmingham and the Black Country*. Hemel Hempstead: Allen and Unwin.

Almeida, P., and B. Kogut. 1997. 'The exploration of technological diversity and the geographic localization of innovation'. *Small Business Economics* 9: 21–31.

Altenburg, T., and J. Meyer-Stamer. 1999. 'How to promote clusters: policy experiences from Latin America'. *World Development* 27: 1693–1713.

Amable, B., R. Barré, and R. Boyer. 1997. *Les Systèmes d'Innovation à l'Ere de la Globalisation*. Paris: Economica.

Amin, A., ed. 1994. *Post-Fordism: A Reader*. Oxford: Blackwell.

—— and P. Cohendet. 2004. *Architectures of Knowledge: Firms, Capabilities, and Communities*. Oxford: Oxford University Press.

—— and K. Robins. 1990. 'The reemergence of regional economies— the mythical geography of flexible accumulation'. *Environment and Planning D: Society and Space* 8: 7–34.

Amin, S. 1973. *Le Développement Inégal; Essai sur les Formations Sociales du Capitalisme Périphérique*. Paris: Les Éditions de minuit.

Amsden, A. H. 1996. 'A strategic policy approach to government intervention in late industrialization'. In *Road Maps to Prosperity: Essays on Growth and Development*, ed. A. Solimano, 119–41. Ann Arbor: University of Michigan Press.

—— 1997. 'Bringing production back in—understanding government's economic role in late industrialization'. *World Development* 25: 469–80.

Anselin, L., A. Varga, and Z. Acs. 1997. 'Local geographic spillovers between university research and high-technology innovations'. *Journal of Urban Economics* 42: 422–48.

Antonelli, C. 2003. 'Knowledge complementarity and fungibility: implications for regional strategy'. *Regional Studies* 37: 595–606.

—— and M. Calderini. 1999. 'The dynamics of localized technological change'. In *The Organization of Economic Innovation in Europe*, eds. A. Gambardelli and F. Malerba, 158–76. Cambridge: Cambridge University Press.

Archibugi, D., J. Howells, and J. Michie. 1999. 'Innovation systems in a global economy'. *Technology Analysis and Strategic Management* 11: 527–39.

Armstrong, W., and T. G. McGee. 1985. *Theatres of Accumulation: Studies in Asian and Latin American Urbanization*. London; New York: Methuen.

Arrow, K. J. 1962. 'Economic welfare and the allocation of resources for invention'. In *The Rate and Direction of Inventive Activity*, 609–26. Princeton: Princeton University Press.

Arthur, B. 1990. 'Silicon Valley locational clusters: when do increasing returns imply monopoly?' *Mathematical Social Science* 19: 235–251.

Assimakopoulos, D., S. Everton, and K. Tsutsui. 2003. 'The semiconductor community in the Silicon Valley: a network analysis of the SEMI genealogy chart (1947–1986)'. *International Journal of Technology Management* 25: 181–99.

Audirac, I. 2003. 'Information-age landscapes outside the developed world—Bangalore, India, and Guadalajara, Mexico'. *Journal of the American Planning Association* 69: 16–23.

Audretsch, D. B. 2002. 'The innovative advantage of US cities'. *European Planning Studies* 10: 165–76.

—— 2003. 'Managing knowledge spillovers: the role of geographic proximity'. In *Geography and Strategy*, eds. J. A. C. Brown and O. Sorenson, 23–48. Oxford: JAI.

—— and M. P. Feldman. 1996. 'R&D spillovers and the geography of innovation and production'. *American Economic Review* 86: 630–40.

Aydalot, P. 1986. 'Trajectoires technologiques et milieux innovateurs'. In *Milieux Innovateurs en Europe*, ed. P. Aydalot, 345–61. Paris: GREMI.

Babbage, C. 1832. *On the Economy of Machinery and Manufactures*. London: C. Knight.

Bagnasco, A. 1977. *Tre Italie: La Problematica Territoriale dello Sviluppo Italiano*. Bologna: Il Mulino.

Baptista, R., and P. Swann. 1998. 'Do firms in clusters innovate more?' *Research Policy* 27: 525–40.

Barnes, B., D. Bloor, and J. Henry. 1996. *Scientific Knowledge: A Sociological Analysis*. London: Athlone.

Barro, R. J. 1997. *Determinants of Economic Growth: A Cross-Country Empirical Study, Lionel Robbins lectures*. Cambridge, Mass.: The MIT Press.

Baud, I. S. A., and G. A. de Bruijne, eds. 1993. *Gender, Small-Scale Industry and Development Policy*. London: IT Publications.

Baumol, W. J. 2002. *The Free-Market Innovation Machine: Analyzing the Growth Miracle of Capitalism*. Princeton: Princeton University Press.

Beaudry, C., and S. Breschi. 2003. 'Are firms in clusters really more innovative?' *Economics of Innovation and New Technology* 12: 325–42.

Becattini, G. 1978. 'The development of light industry in Tuscany'. *Economic Notes* 2–3: 53–72.

Becker, C. M., J. G. Williamson, and E. S. Mills. 1992. *Indian Urbanization and Economic Growth since 1960, The Johns Hopkins Studies in Development*. Baltimore: Johns Hopkins University Press.

Becker, H. S. 1982. *Art Worlds*. Berkeley: University of California Press.

Beesley, M. 1955. 'The birth and death of industrial establishments: experience of the West Midlands conurbation'. *Journal of Industrial Economics* 4: 45–61.

Bellak, C., and J. Cantwell. 1998. 'Globalization tendencies relevant for latecomers'. In *Latecomers in the Global Economy*, eds. M. Storper, S. B. Thomadikis, and L. J. Tsipouri, 40–75. London: Routledge.

Benner, C. 2003. 'Learning communities in a learning region: the soft infrastructure of cross-firm learning networks in Silicon Valley'. *Environment and Planning A* 35: 1809–30.

Bianchi, P. 1992. 'Levels of policy and the nature of post-fordist competition'. In *Pathways to Industrialization and Regional Development*, eds. M. Storper and A. J. Scott, 303–15. London: Routledge.

Bielby, W. T., and D. D. Bielby. 1999. 'Organizational mediation of project-based labor markets: talent agencies and the careers of screenwriters'. *American Sociological Review* 64: 64–85.

Blair, H., S. Grey, and K. Randle. 2001. 'Working in film: employment in a project-based industry'. *Personnel Review* 30: 170–85.

Bloom, D. E., and J. D. Sachs. 1998. 'Geography, demography, and economic growth in Africa'. *Brookings Papers on Economic Activity* (2): 207–95.

Bordwell, D., J. Staiger, and K. Thompson. 1985. *The Classical Hollywood Cinema: Film Style and Mode of Production to 1960*. New York: Columbia University Press.

Boschma, R. A., and J. G. Lambooy. 1999. 'Evolutionary economics and economic geography'. *Journal of Evolutionary Economics* 9: 411–29.

Boudeville, J. R. 1966. *Problems of Regional Economic Planning*. Edinburgh: Edinburgh University Press.

Bourdieu, P. 1972. *Esquisse d'une Théorie de la Pratique*. Geneva: Librairie Droz.

Braverman, H. 1974. *Labor and Monopoly Capital: The Degradation of Work in the Twentieth Century*. New York: Monthly Review Press.

Breschi, S. 1999. 'Spatial patterns of innovation: evidence from patent data'. In *Organization of Economic Innovation in Europe*, eds. A. Gambardella and F. Malerba, 71–102. Cambridge: Cambridge University Press.

—— and F. Malerba. 2001. 'The geography of innovation and economic clustering: some introductory notes'. *Industrial and Corporate Change* 10: 817–33.

Broadberry, S., and A. Marrison. 2002. 'External economies of scale in the Lancashire cotton industry, 1900–1950'. *Economic History Review* 55: 51–77.

Brown, A., J. O'Connor, and S. Cohen. 2000. 'Local music policies within a global music industry: cultural quarters in Manchester and Sheffield'. *Geoforum*: 437–51.

Brown, J. S., and P. Duguid. 2000*a*. 'Mysteries of the region: knowledge dynamics in Silicon Valley'. In *The Silicon Valley Edge*, eds. C. M. Lee, W. F. Miller, M. G. Hancock, and H. S. Rowen, 16–39. Stanford, Calif.: Stanford University Press.

—— —— 2000*b*. *The Social Life of Information*. Boston: Harvard Business School Press.

Brusco, S. 1982. 'The Emilian model: productive decentralization and social integration'. *Cambridge Journal of Economics* 6: 167–80.

Buckley, P. J., and P. N. Ghauri. 2004. 'Globalisation, economic geography and the strategy of multinational enterprises'. *Journal of International Business Studies* 35: 81–98.

Bunnell, T. G. 2002. 'Multimedia utopia? A geographical critique of high-tech development in Malaysia's Multimedia Supercorridor'. *Antipode* 34: 265–95.

—— and N. M. Coe. 2001. 'Spaces and scales of innovation'. *Progress in Human Geography* 25: 569–89.

Cadène, P. 1998. 'Network specialists, industrial clusters and the integration of space in India'. In *Decentralized Production in India: Industrial Districts, Flexible Specialization, and Employment*, eds. P. Cadène and M. Holmström, 139–65. New Delhi: Sage Publications.

Cairncross, F. 1997. *The Death of Distance: How the Communications Revolution Will Change our Lives*. Boston: Harvard Business School Press.

Camagni, R. P. 1995. 'The concept of innovative milieu and its relevance for public policies in European lagging regions'. *Papers in Regional Science* 74: 317–40.

Campbell, T. 2001. 'Innovation and risk-taking: urban governance in Latin America'. In *Global City-Regions: Trends, Theory, Policy*, ed. A. J. Scott, 214–36. Oxford: Oxford University Press.

Cantwell, J., and S. Iamarino. 2002. 'The technological relationships between indigenous firms and foreign-owned MNCs in the European regions'. In *Industrial Location Economics*, ed. P. McCann. Cheltenham: Edward Elgar.

—— and O. Janne. 1999. 'Technological globalisation and innovative centres: the role of corporate technology leadership and locational hierarchy'. *Research Policy* 28: 119–44.

Capello, R. 2002. 'Entrepreneurship and spatial externalities: theory and measurement'. *Annals of Regional Science* 36: 387–402.

Carlino, G. A. 1979. 'Increasing returns to scale in metropolitan manufacturing'. *Journal of Regional Science* 19: 363–73.

Carlton, D. W. 1979. 'Vertical integration in competitive markets under uncertainty'. *Journal of Industrial Economics* 27: 189–209.

Casson, M. 1982. *The Entrepreneur: An Economic Theory*. Cheltenham: Edward Elgar.

Castells, M., and P. Hall. 1994. *Technopoles of the World: The Making of Twenty-First Century Industrial Complexes*. London: Routledge.

Caves, R. E. 2000. *Creative Industries: Contracts between Art and Commerce*. Cambridge, Mass.: Harvard University Press.

Cawthorne, P. 1993. 'The labour process under amoebic capitalism—a case study of the garment industry in a south Indian town'. In

Gender, Small-Scale Industry and Development Policy, eds. I. S. A. Baud and G. A. De Bruijne, 47–75. London: Intermediate Technology Publications.

Cawthorne, P. 1995. 'Of networks and markets: the rise and rise of a south Indian town: the example of Tiruppur's cotton knitwear industry'. *World Development* 23: 43–57.

Chacar, A. S., and M. B. Lieberman. 2003. 'Organizing for technological innovation in the US pharmaceutical industry'. In *Geography and Strategy*, eds. J. A. C. Baum and O. Sorenson, 317–40. Kidlington: JAI.

Chamberlin, E. 1933. *The Theory of Monopolistic Competition*. Cambridge, Mass.: Harvard University Press.

Chapman, S. J., and T. S. Ashton. 1914. 'The sizes of businesses, mainly in the textile industries'. *Journal of the Royal Statistical Society* 77: 469–555.

Chari, S. 2000. 'The agrarian origins of the knitwear industrial cluster in Tiruppur, India'. *World Development* 28: 579–99.

Chen, Y. 1996. 'Impact of regional factors on productivity in China'. *Journal of Regional Science* 36: 417–36.

Christerson, B., and C. Lever-Tracy. 1997. 'The Third China? Emerging industrial districts in rural China'. *International Journal of Urban and Regional Research* 21: 569–88.

Clark, G. L., and W. B. Kim, eds. 1995. *Asian NIEs and the Global Economy: Industrial Restructuring and Corporate Strategy in the 1990s*. Baltimore: The Johns Hopkins University Press.

Coase, R. H. 1937. 'The nature of the firm'. *Economica* 4: 386–405.

Cohendet, P., F. Kern, B. Mehmanpzir, and F. Munier. 1999. 'Knowledge coordination, competence creation and integrated networks in globalized firms'. *Cambridge Journal of Economics* 23: 225–41.

Cooke, P. 2002. *Knowledge Economies: Clusters, Learning and Cooperative Advantage*. London: Routledge.

—— and K. Morgan. 1994. 'The regional innovation system in Baden-Württemberg'. *International Journal of Technology Management* 9: 394–429.

—— —— 1998. *The Associational Economy: Firms, Regions, and Innovation*. Oxford: Oxford University Press.

Cooper, A., and T. Folta. 2000. 'Entrepreneurship in high-technology clusters'. In *The Blackwell Handbook of Entrepreneurship*, eds. D. L. Sexton and H. Landström, 348–67. Oxford: Blackwell.

Corbridge, S. 1986. *Capitalist World Development: A Critique of Radical Development Geography*. Basingstoke: Macmillan.

Crane, D. 1992. *The Production of Culture: Media and the Urban Arts*. Newbury Park, Calif.: Sage.

Creamer, D. B. 1935. *Is Industry Decentralizing? A Statistical Analysis of Locational Changes in Manufacturing Employment, 1899–1933*. Philadelphia: University of Pennsylvania Press.

Crewe, L. 2004. 'Unravelling fashion's commodity chains'. In *Geographies of Commodity Chains*, eds. A. Hughes and S. Reimer, 195–214. London: Routledge.

Cumbers, A., D. Mackinnon, and K. Chapman. 2003. 'Innovation, collaboration and learning in industrial clusters: a study of SMEs in the Aberdeen oil complex'. *Environment and Planning A* 35: 1689–706.

David, P. A. 1985. 'Clio and the economics of QWERTY'. *American Economic Review* 75: 332–7.

—— and D. Foray. 2002. 'An introduction to the economy of the knowledge society'. *International Social Science Journal* 54: 9–23.

Dawson, J. 1992. 'The relevance of the flexible specialisation paradigm for small-scale industrial restructuring in Ghana'. *IDS Bulletin* 23 (3): 34–8.

de la Mothe, J., and G. Paquet. 1998. 'Local and regional systems of innovation as learning socio-economies'. In *Local and Regional Systems of Innovation*, eds. J. de la Mothe and G. Paquet, 1–16. Dordrecht: Kluwer.

de Vet, J. M., and A. J. Scott. 1992. 'The Southern California medical device industry: innovation, new firm formation, and location'. *Research Policy* 21 (2): 145–61.

Dignard, L., and J. Havet, Eds. 1995. *Women in Micro- and Small-Scale Enterprise Development*. Boulder, Colo.: Westview Press.

Domanski, R. 2001. *The Innovative City*. Poznan: Poznan University of Economics.

Dosi, G. 1982. 'Technological paradigms and technological trajectories'. *Research Policy* 11: 147–62.

Douglass, M. 2001. 'Intercity competition and the question of economic resilience: globalization and crisis in Asia'. In *Global City-Regions:*

Trends, Theory, Policy, ed. A. J. Scott, 236–62. Oxford: Oxford University Press.

Drake, G. 2003. ' "This place gives me space": place and creativity in the creative industries'. *Geoforum* 34: 511–24.

Dunning, J. H. 1993. *Multinational Enterprises and the Global Economy*. Wokingham: Addison-Wesley.

Durkheim, E. 1893. *De la division du travail social: Etude sur l'organisation des sociétés supérieures*. Paris: Alcan.

Edquist, C. 1997. 'Systems of innovation approaches—their emergence and characteristics'. In *Systems of Innovation: Technologies, Institutions and Organizations*, ed. C. Edquist. London: Pinter.

Elfring, T., and W. Hulsink. 2003. 'Networks in entrepreneurship: the case of high-technology firms. *Small Business Economics* 21: 409–22.

Emmanuel, A. 1969. *L'Échange Inégal*. Paris: Maspéro.

Ernst, D., and L. Kim. 2002. 'Global production networks, knowledge diffusion, and local capability formation'. *Research Policy* 31: 1417–29.

Fan, C. C., and A. J. Scott. 2003. 'Industrial agglomeration and development: a survey of spatial economic issues in East Asia and a statistical analysis of Chinese regions'. *Economic Geography* 79: 295–319.

Feldman, M. P. 1994. *The Geography of Innovation*. Dordrecht: Kluwer.

—— and D. B. Audretsch. 1999. 'Innovation in cities: science-based diversity, specialization and localized competition'. *European Economic Review* 43: 409–29.

—— and R. Florida. 1994. 'The geographic sources of innovation: technological infrastructure and product innovation in the United States'. *Annals of the Association of American Geographers* 84: 210–29.

Feller, I. 2002. 'Impacts of research universities on technological innovation in industry: evidence from engineering research centers'. *Research Policy* 31: 457–74.

Fine, B. 2003. 'New growth theory'. In *Rethinking Development Economics*, ed. H. J. Chang, 201–17. London: Anthem Press.

Fischer, M. M., J. R. Diez, and F. Snickars. 2001. *Metropolitan Innovation Systems: Theory and Evidence for Three Metropolitan Regions in Europe*. Berlin: Springer.

Florence, P. S. 1948. *Investment, Location and Size of Plant*. Cambridge: Cambridge University Press.

Florida, R. 1995. 'Toward the learning region'. *Futures* 27: 527–36.

―― 2002. *The Rise of the Creative Class*. New York: Basic Books.

―― and M. Kenney. 1988. 'Venture capital, high technology, and regional development'. *Regional Studies* 22: 33–48.

Foray, D., and B. A. Lundvall. 1996. 'The knowledge-based economy: from the economics of knowledge to the learning economy'. In *Employment and Growth in the Knowledge-Based Economy*, eds. D. Foray and B. A. Lundvall. Paris: OECD.

―― and W. E. Steinmuller. 2003. 'The economics of knowledge production by inscription'. *Industrial and Corporate Change* 12: 299–319.

Fornahl, D. 2003. 'Entrepreneurial activities in a regional context'. In *Cooperation, Networks and Institutions in Regional Innovation Systems*, eds. D. Fornahl and T. Brenner, 38–57. Cheltenham: Edward Elgar.

Frank, A. G. 1978. *Dependent Accumulation and Underdevelopment*. London: Macmillan.

Freeman, C. 1987. *Technology Policy and Economic Performance: Lessons from Japan*. London: Pinter.

―― 1995. 'The national system of innovation in historical perspective'. *Cambridge Journal of Economics* 19: 5–24.

Friedmann, G. 1956. *Le Travail en miettes*. Paris: Gallimard.

Fröbel, F., J. Heinrichs, and O. Kreye. 1980. *The New International Division of Labor*. Cambridge: Cambridge University Press.

Gereffi, G. 1995. 'Global production systems and third world development'. In *Global Change, Regional Response: The New International Context of Development*, ed. B. Stallings, 100–42. Cambridge: Cambridge University Press.

―― and M. Korzeniewicz, eds. 1994. *Commodity Chains and Global Capitalism*. Westport, Conn.: Greenwood Press.

Gerschenkron, A. 1962. *Economic Backwardness in Historical Perspective: A Book of Essays*. Cambridge, Mass.: Belknap Press of Harvard University Press.

Gertler, M. S. 1995. 'Being there: proximity, organization and culture in the development and adoption of advanced manufacturing technologies'. *Economic Geography* 71: 1–26.

―― 2003. 'Tacit knowledge and the economic geography of context, or, the undefinable tacitness of being (there)'. *Journal of Economic Geography* 3: 75–99.

Gibson, C. 2003. 'Cultures at work: why culture matters in research on the cultural industries'. *Social and Cultural Geography* 4: 201–15.

Giddens, A. 1984. *The Constitution of Society: Outline of a Theory of Structuration*. Berkeley: University of California Press.

Glaeser, E. L., H. Kallal, J. Schienkman, and A. Shleifer. 1992. 'Growth in cities'. *Journal of Political Economy* 100: 1126–52.

Grabher, G. 2001. 'Ecologies of creativity: the village, the group, and the heterarchic organization of the British advertising industry'. *Environment and Planning A* 33: 351–74.

—— 2002. 'Cool projects, boring institutions: temporary collaboration in social context'. *Regional Studies* 36: 205–14.

Granovetter, M. S. 1973. 'The strength of weak ties'. American Journal of Sociology (78): 1360–80.

Grant, R. 2001. 'Liberalization policies and foreign companies in Accra, Ghana'. *Environment and Planning, A* 33: 997–1014.

Griliches, Z. 1990. 'Patent statistics as economic indicators: a survey'. *Journal of Economic Literature* 28: 1661–707.

Grossetti, M. 1995. *Science, Industrie et Territoire*. Toulouse: Presses Universitaires du Mirail.

Haig, R. M. 1927. *Major Economic Factors in Metropolitan Growth and Arrangement*. New York: Regional Plan of New York and its Environs.

Hall, P. 1962. *The Industries of London since 1861*. London: Hutchinson.

—— 1998. *Cities in Civilization*. New York: Pantheon.

Harcourt, G. C. 1972. *Some Cambridge Controversies in the Theory of Capital*. Cambridge: Cambridge University Press.

Harvey, D. 1989. 'From managerialism to entrepreneurialism—the transformation in urban governance in late capitalism'. *Geografiska Annaler Series B—Human Geography* 71: 3–17.

He, C. F. 2003. 'Location of foreign manufacturers in China: agglomeratin economies and country of origin effects'. *Papers in Regional Science* 82: 351–72.

Heckscher, E. F., and B. Ohlin. 1991. *Heckscher-Ohlin Trade Theory*, trans. H. Flam and M. J. Flanders. Cambridge, Mass.: MIT Press.

Henderson, J. V. 1986. 'Efficiency of resource usage and city size'. *Journal of Urban Economics* 19: 47–70.

—— 1988. *Urban Development: Theory, Fact, and Illusion*. New York: Oxford University Press.

—— and A. Juncoro. 1996. 'Industrial centralization in Indonesia'. *World Bank Economic Review* 10: 513–40.

—— Z. Shalizi, and A. J. Venables. 2001. 'Geography and development'. *Journal of Economic Geography* 1: 81–105.

Henderson, J. W. 1989. *The Globalisation of High Technology Production: Society, Space, and Semiconductors in the Restructuring of the Modern World*. London; New York: Routledge.

Hennion, A. 1981. *Les Professionels du Disque: Une Sociologie des Variétés*. Paris: Editions A. M. Métailié.

Hirsch, P. M. 1972. 'Processing fads and fashions: an organization-set analysis of cultural industry systems'. *American Journal of Sociology* 77: 639–59.

Hirschman, A. O. 1958. *The Strategy of Economic Development, Yale Studies in Economics 10*. New Haven: Yale University Press.

Hoover, E. M. 1937. *Location Theory and the Shoe and Leather Industries*. Cambridge, Mass.: Harvard University Press.

—— and R. Vernon. 1959. *Anatomy of a Metropolis*. Cambridge, Mass.: Harvard University Press.

Howells, J. 1999. 'Regional systems of innovation?' In *Innovation Policy in a Global Economy*, eds. D. Archibugi and J. Michie, 67–93. Cambridge: Cambridge University Press.

Humphrey, J., and H. Schmitz. 2002. 'How does insertion in global value chains affect upgrading in industrial clusters?' *World Development* 36: 1017–27.

Jacobs, J. 1961. *The Death and Life of Great American Cities*. New York: Vintage Books.

———. 1969. *The Economy of Cities*. New York: Random House.

Jaffe, A. B., M. Trajtenberg, and R. Henderson. 1993. 'Geographic localization of knowledge spillovers as evidenced by patent citations'. *Quarterly Journal of Economics* 108: 577–98.

Jeffcut, P., and A. C. Pratt. 2002. 'Managing creativity in the cultural industries'. *Creativity and Innovation Management* 11: 225–33.

Kaldor, N. 1970. 'The case for regional policies'. *Scottish Journal of Political Economy* 17: 337–48.

Kattuman, P. A. 1998. 'The role of history in the transition to an industrial district: the case of the Indian bicycle industry'. In *Decentralized Production in India: Industrial Districts, Flexible Specialization, and Employment*, eds. P. Cadène and M. Holmström, 230–50. Thousand Oaks, Calif.: Sage Publications.

Kaufmann, A., P. Lehner, and F. Tödtling. 2003. 'Effects of the internet on the spatial structure of innovation networks'. *Information, Economics and Policy* 15: 402–24.

Kawashima, T. 1975. 'Urban agglomeration economies in manufacturing industries'. *Papers of the Regional Science Association* 34: 157–75.

Kealy, E. R. 1979. 'From craft to art: the case of sound mixers and popular music'. *Sociology of Work and Occupations* 6: 3–29.

Kessler, J. A. 1999. 'The North American Free Trade Agreement, emerging apparel production networks and industrial upgrading: the Southern California/Mexico connection'. *Review of International Political Economy* 6: 565–608.

Klink, J. J. 2001. *A Cidade-Região: Regionalismo e Reestructuração no Grande ABC Paulista*. Rio de Janeiro: De Paulo Editora.

Knorringa, P. 1996. *Economics of Collaboration: Indian Shoemakers between Market and Hierarchy*. Thousand Oaks, Calif. Sage Publications.

—— 1999. 'Agra: an old cluster facing new competition'. *World Development* 27: 1587–604.

Krueger, A. O. 2000. 'Government failures in development'. In *Modern Political Economy and Latin America*, eds. J. Frieden, M. Pastor, and M. Tomz. Boulder, Colo.: Westview.

Krugman, P. 1991. *Geography and Trade*. Leuven, Belgium: Leuven University Press.

—— 1996. *Development, Geography, and Economic Theory*. Cambridge, Mass.: The MIT Press.

Lal, D. 1983. *The Poverty of 'Development Economics'*. London: Institute of Economic Affairs.

Lall, S. 1980. 'Vertical inter-firm linkages in LDCs: an empirical study'. *Oxford Bulletin of Economics and Statistics* 42: 203–26.

—— 1990. *Building Industrial Competitiveness in Developing Countries, Development Centre Studies*. Paris, France: Development Centre of the Organisation for Economic Co-operation and Development.

—— Z. Shalizi, and U. Deichmann. 2004. 'Agglomeration economies and productivity in Indian industry'. *Journal of Development Economics* 73: 643–73.

Lamoreaux, N. R., and K. L. Sokoloff. 2000. 'The geography of invention in the American glass industry: 1870–1925'. *Journal of Economic History* 60: 700–29.

Lampard, E. E. 1955. 'The history of cities in economically advanced areas'. *Economic Development and Cultural Change* 3: 81–102.

Landry, C. 2000. *The Creative City: A Toolkit for Urban Innovators*. London: Earthscan.

Latour, B., and S. Woolgar. 1979. *Laboratory Life: The Social Construction of Scientific Facts*. Beverly Hills, Calif.: Sage.

Lawson, C. 1999. 'Towards a competence theory of the region'. *Cambridge Journal of Economics* 23: 151–66.

Lawson, V. 1995. 'Beyond the firm: restructuring gender divisions of labor in Quito's garment industry under austerity'. *Environment and Planning A* 13: 415–44.

Lazonick, W. 1983. 'Industrial organization and technical change: the decline of the British cotton textile industry'. *Business History Review* 57: 195–236.

Leamer, E. E., and M. Storper. 2001. 'The economic geography of the internet era'. *Journal of International Business Studies* 32: 641–65.

Lee, J. 1997. 'The maturation and growth of infant industries'. *World Development* 25: 1271–81.

Lee, Y. J., and Zang, H. 1998. 'Urbanization and regional productivity in Korean manufacturing'. *Urban Studies* 35: 2085–99.

Leibenstein, H. 1954. *A Theory of Economic-Demographic Development*. Princeton: Princeton University Press.

Leijonhufvud, A. 1986. 'Capitalism and the factory system'. In *Economics as a Process: Essays in the New Institutional Economics*, ed. R. N. Langlois, 203–23. New York: Cambridge University Press.

Levine, R., and D. Renelt. 1992. 'A sensitivity analysis of cross-country growth regressions'. *American Economic Review* 82: 942–63.

Lewis, W. A. 1954. 'Economic development with unlimited supplies of labour'. *Manchester School* (May): 139–91.

Leydesdorff, L., and H. Etzkowitz. 1997. 'A triple helix of university-industry-government relations'. In *Universities and the Global Knowledge Economy: A Triple Helix of University-Industry-Government Relations*, eds. H. Etzkowitz and L. Leydesdorff, 155–62. London: Pinter.

Lichtenburg, R. M. 1960. *One-Tenth of a Nation*. Cambridge, Mass.: Harvard University Press.

Lipietz, A. 1986. 'New tendencies in the international division of labor: regimes of accumulation and modes of social regulation'. In

Production, Work, Territory: The Anatomy of Industrial Capitalism, eds. A. J. Scott and M. Storper, 16–40. Boston: Allen and Unwin.

Lipton, M. 1977. *Why Poor People Stay Poor: A Study of Urban Bias in World Development*. London: Temple Smith.

Lissoni, F. 2001. 'Knowledge codification and the geography of innovation: the case of the Brescia mechanical cluster'. *Research Policy* 30: 1479–500.

List, F. 1977; 1841. *National System of Political Economy*. Fairfield, NJ: A. M. Kelley.

Livingstone, D. N. 1995. 'The spaces of knowledge: contributions towards a historical geography of science'. *Environment and Planning D: Society and Space* 13: 5–34.

Lucas, R. E. 1988. 'On the mechanics of economic development'. *Journal of Monetary Economics* 22: 3–42.

—— 1993. 'Making a miracle'. *Econometrica* 61: 251–72.

Luger, M. I., and H. A. Goldstein. 1991. *Technology in the Garden: Research Parks and Regional Economic Development*. Chapel Hill, NC: University of North Carolina Press.

Lundvall, B. A. 1988. 'Innovation as an interactive process: from user-producer interaction to the national system of innovation'. In *Technical Change and Economic Theory*, eds. G. Dosi, C. Freeman, R. Nelson, G. Silverberg, and L. Soete, 349–69. London: Pinter.

—— and B. Johnson. 1994. 'The learning economy'. *Journal of Industrial Studies* 1: 23–42.

McAleese, D., and D. McDonald. 1978. 'Employment growth and the development of linkages in foreign-owned and domestic manufacturing enterprises'. *Oxford Bulletin of Statistics* 40: 321–9.

McCormick, D. 1999. 'African enterprise clusters and industrialization: theory and reality'. *World Development* 27: 1531–51.

McLaughlin, G. E., and S. Robock. 1949. *Why Industry Moves South: A Study of Factors Influencing the Recent Location of Manufacturing Plants in the South*. Washington: National Planning Association.

Maillat, D., and J. Y. Vasserot. 1986. *Les Milieux Innovateurs: Le Cas de l'Arc Jurassien Suisse*, ed. P. Aydalot, 217–46. Paris: GREMI.

Malecki, E. J. 1991. *Technology and Economic Development*. London: Longman.

Malmberg, A., and P. Maskell. 2002. 'The elusive concept of localization economies: towards a knowledge-based theory of spatial clustering'. *Environment and Planning A* 34: 429–49.

Mansfield, E. 1968. *The Economics of Technological Change*. New York: Norton.

Marshall, A. 1890. *Principles of Economics*. London, New York: Macmillan.

—— 1919. *Industry and Trade: A Study of Industrial Technique and Business Organization*. London: Macmillan.

Marx, K., and F. Engels. 1947. *The German Ideology*. New York: International Publishers.

Maskell, P., and A. Malmberg. 1999. 'Localised learning and industrial competitiveness'. *Cambridge Journal of Economics* 23: 167–85.

—— and G. Törnquist. 2003. 'The role of universities in the learning region'. In *Economic Geography of Higher Education: Knowledge Infrastructure and Learning Regions*, eds. R. Rutten, F. Boekema, and E. Kuijpers, 129–44. London: Routledge.

Massey, D. 1984. *Spatial Divisions of Labor: Social Structures and the Geography of Production*. New York: Methuen.

Menger, P. M. 1993. 'L'hégémonie parisienne: économie et politique de la gravitation artistique'. *Annales: Economies, Sociétés, Civilisations* (6): 1565–600.

Meyer-Stamer, J. 1998. 'Path dependence in regional development: persistence and change in three industrial clusters in Santa Catarina, Brazil'. *World Development* 26: 1495–511.

Mies, M. 1998. *Patriarchy and Accumulation on a World Scale: Women in the International Division of Labor*. London: Zed Books.

Mill, J. S. 1848/1909. *Principles of Political Economy*. London: Longmans, Green.

Mills, E. S., and C. M. Becker. 1986. *Studies in Indian Urban Development*. New York: Oxford University Press.

Mitra, A. 2000. 'Total factor productivity growth and urbanization economies: a case of Indian industries'. *Review of Urban and Regional Development Studies* 12: 97–108.

Molotch, H. 1996. 'LA as design product: how art works in a regional economy'. In *The City: Los Angeles and Urban Theory at the End of the Twentieth Century*, eds. A. J. Scott and E. W. Soja, 225–275. Berkeley and Los Angeles: University of California Press.

Montgomery, S. S., and M. D. Robinson. 1993. 'Visual artists in New York: what's special about person and place?' *Journal of Cultural Economics* 17: 17–39.

Morgan, K. 1997. 'The learning region: institutions, innovation and regional renewal'. *Regional Studies* 31: 491–503.

—— 2004. 'The exaggerated death of geography: learning, proximity and territorial innovation systems'. *Journal of Economic Geography* 4: 3–21.

Moroney, J. R., and J. M. Walker. 1966. 'A regional test of the Heckscher-Ohlin hypothesis'. *Journal of Political Economy* 74: 573–86.

Morris, A. S., and S. Lowder. 1992. 'Flexible specialization: the application of theory in a poor-country context: Leon, Mexico'. *International Journal of Urban and Regional Research* 16: 190–201.

Morshidi, S., and G. Suriati. 1999. *Globalisation of Economic Activity and Third World Cities: A Case Study of Kuala Lumpur*. Kuala Lumpur: Utusan Publications & Distributors.

Moulaert, F., and F. Sekia. 2003. 'Territorial innovation models: a critical survey'. *Regional Studies* 37: 289–302.

Mulkay, M. J. 1972. *The Social Process of Innovation: A Study in the Sociology of Science*. London: Macmillan.

Murphy, K. M., A. Schleifer, and R. W. Vishny. 1989. 'Industrialization and the big push'. *Journal of Political Economy*: 1003–26.

Myrdal, G. 1959. *Economic Theory and Under-developed Regions*. London: Gerald Duckworth & Co.

Nadvi, K. 1999a. 'Shifting ties: social networks in the surgical instrument cluster of Sialkot, Pakistan'. *Development and Change* 30: 141–75.

—— 1999b. 'Collective efficiency and collective failure: the response of the Sialkot surgical instrument cluster to global quality pressures'. *World Development* 27: 1605–26.

Nelson, R. R., ed. 1993. *National Innovation Systems: A Comparative Analysis*. New York: Oxford University Press.

—— and S. G. Winter. 1982. *An Evolutionary Theory of Economic Change*. Cambridge, Mass.: Belknap Press.

Nijkamp, P. 2003. 'Entrepreneurship in a modern network economy'. *Regional Studies* 37: 395–404.

Niosi, J., and T. G. Bas. 2001. 'The competencies of regions—Canada's clusters in biotechnology'. *Small Business Economics* 17: 31–42.

Nonaka, I. 1994. 'A dynamic theory of organizational knowledge creation'. *Organization Science* 5: 14–37.

North, D. C. 1998. 'Institutions, ideology and economic performance'. In *The Revolution in Development Economics*, eds. J. A. Dorn, S. H. Hanke, and A. A. Walters, 95–107. Washington: Cato Institute.

Norton, R. D., and J. Rees. 1979. 'The product cycle and the spatial decentralization of American manufacturing'. *Regional Studies* 13: 141–51.

Noteboom, B. 1999. 'Innovation, learning and industrial organization'. *Cambridge Journal of Economics* 23: 127–50.

Nurske, R. 1959. *Patterns of Trade and Development*. Stockholm: Almqvist & Wiksell.

O'Brien, R. O. 1992. *Global Financial Integration: The End of Geography*. London: Royal Institute of International Affairs.

Ó hUallacháin, B. 1999. 'Patent places: size matters'. *Journal of Regional Science* 39: 613–36.

Oinas, P., and E. J. Malecki. 1999. 'Spatial innovation systems'. In *Making Connections: Technological Learning and Regional Economic Change*, eds. E. J. Malecki and P. Oinas, 7–33. Aldershot: Ashgate.

—— 2002. 'The evolution of technologies in time and space: from national and regional to spatial innovation systems'. *International Regional Science Review* 25: 102–31.

Orsi, F., and B. Coriat. 2003. 'Intellectual property rights, financial markets and innovation: a sustainable configuration?' *Issues in Regulation Theory* (45): 1–4.

Palley, T. I. 2003. 'Export-led growth: evidence of developing country crowding out'. In *Globalisation, Regionalism and Economic Activity*, eds. P. Arestis, M. Baddeley, and J. McCombie, 175–97. Cheltenham: Edward Elgar.

Pan, Z. H., and F. Zhang. 2002. 'Urban productivity in China'. *Urban Studies* 39: 2267–81.

Pathania-Jain, G. 2001. 'Global patents, local partnerships: a value-chain analysis of collaborative strategies of media firms in India'. *Journal of Media Economics* 14: 169–87.

Pavitt, K., and P. Patel. 1991. 'Large firms in the production of the world's technology: an important case of non-globalisation'. *Journal of International Business Studies* 22: 1–21.

Peck, J. 1996. *Work-place: The Social regulation of Labor Markets, Perspectives on Economic Change*. New York: Guilford Press.

Perez, C. 1983. 'Structural change and assimilation of new technologies in the economic and social systems'. *Futures* 15: 357–75.

Perrin, M. 1937. *Saint-Étienne et sa Région Économique: un Type de la Vie Industrielle en France*. Tours: Arrault.

Perroux, F. 1961. *L'Économie du XXe Siècle*. Paris: Presses Universitaires de France.

Peterson, R. A., and D. G. Berger. 1975. 'Cycles in symbol production: the case of popular music'. *American Sociological Review* 40: 158–73.

Pinch, S., and N. Henry. 1999. 'Discursive aspects of technological innovation: the case of the British motor-sport industry'. *Environment and Planning A* 31: 665–82.

—— M. Jenkins, and S. Talman. 2003. 'From industrial districts to knowledge clusters: a model of knowledge dissemination and competitive advantage in industrial agglomerations'. *Journal of Economic Geography* 3: 373–88.

Polanyi, K. 1944. *The Great Transformation*. New York: Farrar and Rinehart.

Polanyi, M. 1966. *The Tacit Dimension*. New York: Doubleday.

Pollard, J. 2004. 'Manufacturing culture in Birmingham's jewelry quarter'. In *Cultural Industries and the Production of Culture*, eds. D. Power and A. J. Scott, 169–87. London: Routledge.

Pollard, S. 1981. *Peaceful Conquest: The Industrialization of Europe, 1760–1970*. Oxford; New York: Oxford University Press.

Porter, M. 1990. *The Competitive Advantage of Nations*. New York: The Free Press.

Powell, W. W., K. W. Koput, and L. Smith-Doerr. 1996. 'Interorganizational collaboration and the locus of innovation: networks of learning in biotechnology'. *Administrative Science Quarterly* 41: 116–45.

Power, D. 2002. 'Cultural industries in Sweden: an assessment of their place in the Swedish economy'. *Economic Geography* 78: 103–27.

Pratt, A. C. 1997. 'The cultural industries production system: a case study of employment change in Britain, 1984–91'. *Environment and Planning A* 29: 1953–74.

Prebisch, R. 1959. 'Commercial policy in the underdeveloped countries'. *American Economic Review, Papers and Proceeding* 44: 251–73.

Rabellotti, R. 1997. *External Economies and Cooperation in Industrial Districts*. London: Macmillan.

—— 1999. 'Recovery of a Mexican cluster: devaluation bonanza or collective efficiency?' *World Development* 27: 1571–85.

—— and H. Schmitz. 1999. 'The internal heterogeneity of industrial districts in Italy, Brazil, and Mexico'. *Regional Studies* 33: 97–108.

Rallet, A., and A. Torre. 1999. 'Is geographical proximity necessary in the innovation networks in the era of global economy?' *GeoJournal* 49: 373–80.

Rantisi, N. 2004. 'The designer in the city and the city in the designer'. In *Cultural Industries and the Production of Culture*, eds. D. Power and A. J. Scott, 91–109. London: Routledge.

Rasiah, R. 1994. 'Flexible production systems and local machine-tool subcontracting: electronics components transnationals in Malaysia'. *Cambridge Journal of Economics* 18: 279–98.

Renaud, B. 1979. *National Urbanization Policies in Developing Countries, World Bank Staff Working Paper; no. 347*. Washington: World Bank.

Ricardo, D. 1817. *Principles of Political Economy and Taxation*. Harmondsworth: Penguin Books (1971 edn).

Richardson, H. W. 1980. 'Polarization reversal in developing countries'. *Papers of the Regional Science Association* 45: 67–85.

—— 1993. 'Efficiency and welfare in LDC mega-cities'. In *Third World Cities: Problems, Policies and Prospects*, eds. A. M. Parnell and J. D. Kasarda, 32–57. Newbury Park: Sage Publications.

Roberts, M. W. 1992. *Export Processing Zones in Jamaica and Mauritius: Evolution of an Export-Oriented Development Model*. San Francisco: Mellen Research University Press.

Robinson, E. A. G. 1931. *The Structure of Competitive Industry*. Cambridge: Cambridhe University Press.

Rodríguez-Pose, A., and M. C. Refolo. 2003. 'The link between local production systems and public university research in Italy'. *Environment and Planning A* 35: 1477–492.

Rodrik, D. 1999. *The New Global Economy and Developing Countries: Making Openness Work*. Washington: Overseas Development Council.

Romanelli, E., and C. B. Schoonhaven. 2001. 'The local origins of new firms'. In *The Entrepreneurship Dynamic: Origins of Entrepreneurship and the Evolution of Industries*, eds. C. B. Schoonhaven and E. Romanelli, 40–67. Stanford, Calif.: Stanford University Press.

Romer, P. M. 1986. 'Increasing returns and long-run growth'. *Journal of Political Economy* 94: 1002–37.

Roost, F. 1998. 'Recreating the city as entertainment center: the media industry's role in transforming Potsdamer Platz and Times Square'. *Journal of Urban Technology* 5: 1–21.

Rosenberg, N. 1982. *Inside the Black Box: Technology and Economics.* Cambridge: Cambridge University Press.

Rosenstein-Rodan, P. 1943. 'Problems of industrialization of Eastern and South-Eastern Europe'. *Economic Journal* 53: 202–11.

Rostow, W. W. 1960. *The Stages of Economic Growth: A Non-Communist Manifesto.* Cambridge: Cambridge University Press.

Russo, M. 1985. 'Technical change and the industrial district: the role of interfirm relations in the growth and transformation of ceramic tile production in Italy'. *Research Policy* 14: 329–43.

Sabel, C. 1993. 'Studied trust: building new forms of cooperation in a volatile economy'. *Social Relations* 46: 1133–71.

Sachs, J. D., and A. M. Warner. 1997. 'Sources of slow growth in African economies'. *Journal of African Economies* 6: 335–76.

Samuelson, P. A. 1948. 'International trade and the equalisation of factor prices'. *Economic Journal* 58: 163–84.

Sandee, H. 1994. 'The impact of technological change on interfirm linkages: a case study of clustered rural small-scale roof tile enterprises in central Java'. In *Flexible Secialization: The Dynamics of Small-Scale Industry in the South*, eds. P. O. Pederson, A. Sverisson, and M. P. van Dijk, 84–96. London: Intermediate Technology Publications.

Santagata, W. 2004. 'Creativity, fashion, and market behavior'. In *Cultural Industries and the Production of Culture*, eds. D. Power and A. J. Scott, 75–90. London: Routledge.

Saxenian, A. 1994. *Regional Advantage: Culture and Competition in Silicon Valley and Route 128.* Cambridge, Ma.: Harvard University Press.

—— 2002. 'Bangalore: The Silicon Valley of Asia?' In *Economic Policy Reforms and the Indian Economy*, ed. A. O. Krueger. Chicago: University of Chicago Press.

Say, J. B. 1803. *Traité d'Economie Politique.* Paris: Déterville.

Sayer, A., and R. Walker. 1992. *The New Social Economy: Reworking the Division of Labor.* Oxford: Blackwell.

Schmitz, H. 1995. 'Small shoemakers and fordist giants: tales of a supercluster'. *World Development* 23: 9–28.

—— 1999*a*. 'Collective efficiency and increasing returns'. *Cambridge Journal of Economics* 23: 465–83.

—— 1999*b*. 'From ascribed to earned trust in exporting clusters'. *Journal of International Economics* 48: 139–50.

—— 2001. 'Local governance and conflict management: reflections on a Brazilian cluster'. In *Global City-Regions: Trends, Theory, Policy*, ed. A. J. Scott, 401–6. Oxford: Oxford University Press.

—— ed. 2004. *Local Enterprises in the Global Economy: Issues of Governance and Upgrading*. Cheltenham: Edward Elgar.

Schumpeter, J. A. 1934. *The Theory of Economic Development*. Cambridge, Mass.: Harvard University Press.

—— 1942. *Capitalism, Socialism and Democracy*. New York: Harper and Row.

Schweizer, T. S. 2003. 'Managing interactions between technological and stylistic innovation in the media industries'. *Technology Analysis and Strategic Management* 15: 19–41.

Scitovsky, T. 1954. 'Two concepts of external economies'. *Journal of Political Economy* 62: 143–51.

Scott, A. J. 1983. 'Industrial organization and the logic of intra-metropolitan location: I. Theoretical considerations'. *Economic Geography* 59: 233–50.

—— 1986. 'High technology industry and territorial development: the rise of the Orange County complex, 1955–1984'. *Urban Geography* 7: 3–45.

—— 1987. 'The semiconductor industry in South-East Asia: organization, location and the international division of labor'. *Regional Studies* 21: 143–60.

—— 1993. *Technopolis: High-Technology Industry and Regional Development in Southern California*. Berkeley: University of California Press.

—— 1994. 'Variations on the theme of agglomeration and growth: the gem and jewelry industry in Los Angeles and Bangkok'. *Geoforum* 25: 249–63.

—— 1998*a*. 'Multimedia and digital visual effects: an emerging local labor market'. *Monthly Labor Review* 121 (3): 30–8.

—— 1998*b*. *Regions and the World Economy: The Coming Shape of Global Production, Competition, and Political Order*. Oxford; New York: Oxford University Press.

Scott, A.J. 1999a. 'The cultural economy: geography and the creative field'. *Culture, Media, and Society* 21: 807–17.

—— 1999b. 'The US recorded music industry: on the relations between organization, location, and creativity in the cultural economy'. *Environment and Planning A* 31: 1965–84.

—— 2000. *The Cultural Economy of Cities: Essays on the Geography of Image-Producing Industries.* London: Sage.

—— 2001a. 'Industrial revitalization in the ABC municipalities of São Paulo: diagnostic analysis and strategic recommendations for a new economy and a new regionalism'. *Regional Development Studies* 7: 1–32.

—— 2001b. 'Introduction'. In *Global City-Regions: Trends, Theory, Policy*, ed. A. J. Scott, 1–8. Oxford: Oxford University Press.

—— 2002a. 'Competitive dynamics of Southern California's clothing industry: the widening global connection and its local ramifications'. *Urban Studies* 39: 1287–1306.

—— 2002b. 'A new map of Hollywood: the production and distribution of American motion pictures'. *Regional Studies* 36: 957–75.

—— 2002c. 'Regional push: the geography of development and growth in low- and middle-income countries'. *Third World Quarterly* 23: 137–61.

—— 2005a. *On Hollywood: The Place, The Industry.* Princeton: Princeton University Press.

—— 2005b. 'The shoe industry of Marikina City, Philippines, a developing-country cluster in crisis'. *Kasarinlan: Philippine Journal of Third World Studies.*

—— and D. P. Angel. 1987. 'The U.S. semiconductor industry: a locational analysis'. *Environment and Planning A*, 19: 875–912.

—— and M. Storper. 1992. 'Regional development reconsidered'. In *Regional Development and Contemporary Industrial Resposne: Extending Flexible Specialisation*, eds. H. Ernste and V. Meier, 3–24. London: Belhaven.

—— —— 2003. 'Regions, globalization, development'. *Regional Studies* 37: 579–93.

Sheard, P. 1983. 'Auto production systems in Japan: organizational and locational features'. *Autralian Geographical Studies* 21: 49–68.

Shefer, D. 1973. 'Localization economies in SMSAs: a production function analysis'. *Journal of Regional Science* 13: 55–64.

Shukla, V. 1988. *Urban Development and Regional Policy in India: An Econometric Analysis.* Bombay: Himalaya Publishing House.

Simmie, J. 2003. 'Innovation and urban regions as national and international nodes for the transfer and sharing of knowledge'. *Regional Studies* 37: 607–20.

—— 2004. 'Innovation and clustering in the globalised international economy'. *Urban Studies* 41: 1095–112.

Singer, H. W. 1950. 'The distribution of gains between investing and borrowing countries'. *American Economic Review, Papers and Proceedings* 40: 473–85.

Smith, A. 1776/1965. *An Inquiry into the Nature and Causes of the Wealth of Nations.* New York: Random House.

Sorenson, O. 2003. 'Social networks and industrial geography'. *Journal of Evolutionary Economics* 13: 513–27.

—— and P. G. Audia. 2000. 'The social structure of entrepreneurial activity: geographic concentration of footwear production in the United States, 1940–1989'. *American Journal of Sociology* 106: 424–61.

Steed, G. P. F. 1971. 'Internal organization, firm integration, and locational change: the Northern Ireland linen complex, 1954–64'. *Economic Geography* 47: 371–83.

Stigler, G. J. 1951. 'The division of labor is limited by the extent of the market'. *Journal of Political Economy* 69: 213–25.

Stiglitz, J. E. 1989. 'Markets, market failures and development'. *American Economic Review* 79: 197–203.

—— 2002. *Globalization and its Discontents.* New York: Norton.

Storper, M. 1985. 'Oligopoly and the product cycle: essentialism in economic geography'. *Economic Geography* 61: 260–82.

—— 1991. *Industrialization, Economic Development, and the Regional Question in the Third World.* London: Pion.

—— 1995. 'The resurgence of regional economies, ten years later: the region as a nexus of untraded interdependencies'. *European Urban and Regional Studies* 2: 191–221.

—— 1996. 'Institutions of the knowledge-based economy'. In *Employment and Growth in the Knowledge-Based Economy*, eds. D. Foray and B. A. Lundvall, 255–86. Paris: OECD.

—— 1997. *The Regional World: Territorial Development in a Global Economy, Perspectives on Economic Change.* New York: Guilford Press.

Storper, M., and S. Christopherson. 1987. 'Flexible specialization and regional industrial agglomerations: the case of the US motion-picture industry'. *Annals of the Association of American Geographers* 77: 260–82.

—— and B. Harrison. 1991. 'Flexibility, hierarchy and regional development: the structure of industrial production systems and their forms of governance in the 1990s'. *Research Policy* 20: 407–22.

—— and A. J. Scott. 1990. 'Work organization and local labor markets in an era of flexible production'. *International Labour Review* 129: 573–91.

—— —— 1995. 'The wealth of regions: market forces and policy imperatives in local and global context'. *Futures* 27: 505–26.

—— and R. Walker. 1989. *The Capitalist Imperative: Territory, Technology, and Industrial Growth*. Oxford: Blackwell.

Stren, R. 2001. 'Local governance and social diversity in the developing world: new challenges for globalizing city-regions'. In *Global City-Regions: Trends, Theory, Policy*, ed. A. J. Scott, 193–213. Oxford: Oxford University Press.

Struyk, R. J., and F. J. James. 1975. *Intrametropolitan Industrial Location: The Pattern and Process of Change*. Lexington, Mass.: Lexington Books.

Sveikauskas, L. 1975. 'The productivity of cities'. *Quarterly Journal of Economics* 89: 393–413.

Sverrison, A. 1997. 'Enterprise network and technological change: aspects of light engineering and metal working in Accra'. In *Enterprise Clusters and Networks in Developing Countries*, eds. M. P. Van Dijk and R. Rabellotti, 169–90. London: Frank Cass.

Sydow, J., and U. Staber. 2002. 'The institutional embeddedness of project networks: the case of content production in German television'. *Regional Studies* 36:215–27.

Tewari, M. 1999. 'Successful adjustment in Indian industry: the case of Ludhiana's woolen knitwear cluster'. *World Development* 27: 1651–71.

Todaro, M. P. 1980. 'Urbanization in developing nations: trends, prospects and policies'. *Journal of Geography* 79: 164–74.

Townroe, P. M., and Keen, D. 1984. 'Polarization reversal in the state of São Paulo, Brazil'. *Regional Studies* 18: 45–54.

Ure, A. 1835. *The Philosophy of Manufactures*. London: Charles Knight.

Uzzi, B., and R. Lancaster. 2003. 'Relational embeddedness and learning: the case of bank loan managers and their clients'. *Management Science* 49: 383–99.

van Dijk, M. P. 1997. 'Small enterprise associations and networks: evidence from Accra'. In *Enterprise Clusters and Networks in Developing Countries*, eds. M. P. van Dijk and R. Rabellotti, 131–54. London: Frank Cass.

—— and R. Rabellotti. 1997. 'Clusters and networks as sources of cooperation and technology diffusion for small enterprises in developing countries'. In *Enterprise Clusters and Networks in Developing Countries*, eds. M.P. van Dijk and R. Rabellotti, 1–10. London: Frank Cass.

Varga, A. 2000. 'Local academic knowledge transfers and the concentration of economic activity'. *Journal of Regional Science* 40: 289–309.

Vernon, R. 1966. 'International investment and international trade in the product cycle'. *Quarterly Journal of Economics* 80: 190–207.

Villarán, F. 1993. 'Small-scale industry efficiency groups in Peru'. In *Small-Scale Firms and Development in Latin America*, ed. B. Späth, 158–95. Geneva: International Institute for Labour Studies.

Visser, E. J. 1999. 'A comparison of clustered and dispersed firms in the small-scale clothing industry of Lima'. *World Development* 27: 1553–70.

von Böhm-Bawerk, E. 1891. *The Positive Theory of Capital*. London: Macmillan.

Von Hippel, E. 1988. *The Sources of Innovation*. New York: Oxford University Press.

Waardenburg, J. G. 1993. 'Small-scale units in the Agra leather footwear industry'. In *Gender, Small-Scale Industry and Development Policy*, eds. I. S. A. Baud and G. A. D. Bruijne, 170–86. London: ITA Publications.

Walcott, S. M. 2002. 'Analyzing an innovative environment: San Diego as a bioscience beachhead'. *Economic Development Quarterly* 16: 99–114.

Waldinger, R., and M. Bozorgmehr. 1996. *Ethnic Los Angeles*. New York: Russell Sage Foundation.

Wallerstein, I. M. 1976. *The Modern World System: Capitalist Agriculture and the Origins of the European World-Economy in the Sixteenth Century*. New York: Academic Press.

Weber, A. 1909. *Reine Theorie des Standorts*. Tübingen: J. C. B. Mohr.

Wenger, E. 1998. *Communities of Practice: Learning, Meaning and Identity*. Cambridge: Cambridge University Press.

Westlund, H., and R. Bolton. 2003. 'Local social capital and entrepreneurship'. *Small Business Economics* 21: 77–113.

Wheaton, W. C., and H. Shishido. 1981. 'Urban concentration, agglomeration economies, and the level of economic development'. *Economic Development and Cultural Change* 30: 17–30.

White, H. C., and C. A. White. 1965. *Canvases and Careers: Institutional Change in the French Painting World*. New York: Wiley.

Williamson, J. G. 1965. 'Regional inequality and the process of national development: a description of the patterns'. *Economic Development and Cultural Change* 13: 3–45.

Williamson, O. E. 1975. *Markets and Hierarchies: Analysis and Antitrust Implications*. New York: The Free Press.

—— 1998. 'Transaction cost economics: how it works; where it is headed'. *De Economist* 146: 23–58.

Wise, M. J. 1949. 'On the evolution of the jewellery and gun quarters in Birmingham'. *Transactions of the Institute of British Geographers* 15: 57–72.

World Bank. 1993. *East Asian Miracle: Economic Growth and Public Policy*. New York: Oxford University Press.

Yates, P. L. 1959. *Forty Years of Foreign Trade: A Statistical Handbook with Special Reference to Primary Products and Under-Developed Countries*. London: Allen and Unwin.

Young, A. 1928. 'Increasing returns and economic progress'. *Economic Journal* 38: 527–42.

Zucker, L. G., M. R. Darby, and J. Armstrong. 1998*a*. 'Geographically localized knowledge: spillovers or markets?' *Economic Inquiry* 36: 65–87.

—— —— and M. B. Brewer. 1998*b*. 'Intellectual human capital and the birth of US biotechnology enterprises'. *American Economic Review* 88: 290–306.

Zukin, S. 1991. *Landscapes of Power: From Detroit to Disney World*. Berkeley: University of California Press.

—— 1995. *The Cultures of Cities*. Oxford: Blackwell.

NOTES

CHAPTER 1

1. These contrasting dimensions of integrated and disintegrated cotton textile firms were one of the stimuli for Stigler's analysis as presented above.

2. Some regional economists in the 1950s and 1960s believed that these trends could be explained in terms of Hecksher–Ohlin processes, and more specifically in terms of a locational shift on the part of labour-intensive industries from high-wage to low-wage locations (cf. Moroney and Walker 1966). The early stages of the process certainly do seem to have been dominated by relatively labour-intensive industries, like textiles and shoes, but closer examination suggests that the early stages actually involved only relatively capital-intensive segments of these industries. After the 1960s, decentralization turned into a virtual free-for-all across all sectors as marked by the steady colonization of peripheral areas by branch plants. But the most labour-intensive segments of all, such as administration and R&D, or high-end production of clothing or furniture, tended to be firmly anchored to central city locations. Thus, and contrary to the views of Moroney and Walker, the core actually remained on balance an important focus of labour-intensive economic activity, whereas industry in much of the periphery was marked by high levels of capital intensity.

CHAPTER 2

1. The preceding commentary admittedly runs counter to the emphasis accorded in much current theory to pure unmediated agency and the social autonomy of the subject. Under the guise of the 'cultural turn' this point of emphasis has recently made great strides in a number of social sciences, most notably for our purposes in economic geography. However, an intellectually vigorous economic geography, it seems to me, needs to ward off this kind of sentimental humanism

(while simultaneously offering due acknowledgement of the significant role of culture in the eventuation of social outcomes), not only
on the basis of the theoretical ideas developed by Bourdieu and
Giddens, but also on the *ad hominem* grounds that if transformations
of existing socio-spatial relations by means of free-floating acts of
volition were on the cards then we would presumably already be
within sight of utopia. As it is, these relations ramify with remarkably
stubborn persistence on the landscape at every scale of geographical
resolution.

2. See Moulaert and Sekia (2003) for a useful review of these ideas and
their interrelations.

3. Schumpeter (1934: p. 93) in one of his more maudlin moments refers
to the motives of the entrepreneur in terms of 'the dream and the will
to found a private kingdom . . . the will to conquer . . . the joy of
creativity'.

4. Contrary to most accounts of the origins of Silicon Valley, there was
nothing truly decisive about the existence of Stanford University at
nearby Palo Alto to Shockley's decision (Scott and Angel 1987). It
might be argued that if Shockley had studied location theory (certainly as it then was) his first choice would more likely have been
Southern California with its burgeoning defence industry.

5. The independents pioneer new styles that first attract audiences in
marginal market niches. As some of these styles become popular, the
majors then bring them aggressively into the mainstream. The scene is
now set for new styles to appear on the fringes of the market. And so on.

6. The garment industries of numerous large cities in both low- and
high-income countries illustrate this point well.

CHAPTER 3

1. Williamson (1975: 21), for example, writes that 'in the beginning
there were markets'.

2. The growth-accounting models for panels of countries that applied
economists such as Barro (1997) and Levine and Renelt (1992) have
proposed are almost entirely silent on the possibility that urban and
regional conditions may influence growth outcomes.

3. See Corbridge (1986) for an extended review and critique of the
corporate imperialism school.

INDEX

Index